TECHNICAL
REPORT

Estimating the Value of Water-Use Efficiency in the Intermountain West

David G. Groves, James Griffin, Sara Hajiamiri

Sponsored by the William and Flora Hewlett Foundation

Environment, Energy, and Economic Development

A RAND INFRASTRUCTURE, SAFETY, AND ENVIRONMENT PROGRAM

The research described in this report was sponsored by the William and Flora Hewlett Foundation and was conducted under the auspices of the Environment, Energy, and Economic Development Program (EEED) within RAND Infrastructure, Safety, and Environment (ISE).

Library of Congress Cataloging-in-Publication Data

Groves, David G.
Estimating the value of water-use efficiency in the Intermountain West / David G. Groves, James Griffin, Sara Hajiamiri.
p. cm.
Includes bibliographical references.
ISBN 978-0-8330-4397-9 (pbk. : alk. paper)
1. Water resources development—California. 2. Water-supply—California. I. Griffin, James. II. Hajiamiri, Sara. III. Title.

TC425.M66C35 2008
363.6'10979—dc22

2007052725

The RAND Corporation is a nonprofit research organization providing objective analysis and effective solutions that address the challenges facing the public and private sectors around the world. RAND's publications do not necessarily reflect the opinions of its research clients and sponsors.

Published 2008 by the RAND Corporation
1776 Main Street, P.O. Box 2138, Santa Monica, CA 90407-2138
1200 South Hayes Street, Arlington, VA 22202-5050
4570 Fifth Avenue, Suite 600, Pittsburgh, PA 15213-2665
RAND URL: http://www.rand.org/
To order RAND documents or to obtain additional information, contact
Distribution Services: Telephone: (310) 451-7002;
Fax: (310) 451-6915; Email: order@rand.org

Preface

This document presents the results of a project funded by the William and Flora Hewlett Foundation to develop and demonstrate new methods for assessing the value of water-use efficiency to water utilities in the western United States. This report proposes an exploratory modeling framework to estimate the value of water use–efficiency programs in the intermountain West and then demonstrates the approach through a case study focused on the Denver, Colorado, region. Due to the limited scope of this study, its results pertaining to the Denver Water service area are not definitive. In particular, Denver Water was not a formal partner in the study, and we used publicly available data for many important components of the models, such as system operation, costs, and environmental impacts. With these limitations, several areas of the analysis, such as environmental benefits, are largely suggestive.

The analysis, however, provides a demonstration of how these tools and methodologies can be utilized to inform efficiency-program planning, should generalize to other rapidly growing regions of the western United States, and should be of interest to water planners and managers throughout the region. Specifically, the methodologies presented in this report can be used to support integrated resource-planning activities by enabling agencies to better value efficiency benefits—particularly those agencies without large planning staffs and budgets. Better valuations of efficiency will enable agencies to make better choices among the many options, including water-use efficiency, for ensuring water-supply reliability in their service areas.

This report complements recent work funded by the Jane and Marc Nathanson Family Foundation to evaluate the benefits of water-use efficiency to commercial-building owners: *Evaluating the Benefits and Costs of Increased Water-Use Efficiency in Commercial Buildings* (Groves, Fischbach, and Hickey, 2007).

The RAND Environment, Energy, and Economic Development Program

This research was conducted under the auspices of the Environment, Energy, and Economic Development Program (EEED) within RAND Infrastructure, Safety, and Environment (ISE). The mission of RAND Infrastructure, Safety, and Environment is to improve the development, operation, use, and protection of society's essential physical assets and natural resources and to enhance the related social assets of safety and security of individuals in transit and in their workplaces and communities. The EEED research portfolio addresses environmental quality and regulation, energy resources and systems, water resources and systems, climate, natural hazards and disasters, and economic development—both domestically and internationally. EEED research is conducted for government, foundations, and the private sector.

Questions or comments about this report should be sent to the project leader, David G. Groves (David_Groves@rand.org). Information about the Environment, Energy, and Economic Development Program is available online (http://www.rand.org/ise/environ). Inquiries about EEED projects should be sent to the following address:

Michael Toman, Director
Environment, Energy and Economic Development Program, ISE
RAND Corporation
1200 South Hayes Street
Arlington, VA 22202-5050
703-413-1100, x5189
Michael_Toman@rand.org

Contents

Figures

Tables

Summary

Increasing water-use efficiency is an important management strategy for western water agencies. Evaluating the cost effectiveness of water-efficiency programs relative to supply-enhancement measures can be difficult, however, because not all the benefits of improved efficiency are easily quantified. Tangible, future benefits, such as avoided costs, not only depend on the details of complex management systems, but can also be strongly influenced by future uncertainties that are difficult to characterize. Other nontangible benefits, such as supply reliability and avoided environmental impacts, are difficult to quantify due to a lack of standardized methodologies; poor data availability; and multiple, competing values over outcomes. Without good estimates of efficiency-program cost effectiveness, it may be difficult to identify appropriate efficiency programs for implementation.

This report utilizes two recently released tools by the California Urban Water Conservation Council—the avoided-cost model (AC model) and the environmental-benefit model (EB model)—to estimate the benefits of water use–efficiency programs in the Denver Water service area. The AC model is customized to reflect the short-run (SR) and long-run (LR) incremental benefits of water-use reduction on Denver Water's three water-collection systems—the South Platte River, the Roberts Tunnel system, and the Moffat system. We then use the methodological approach developed for the EB model to estimate the benefits of water-use efficiency to the environmental and recreational services. Specifically, we evaluate benefits to riparian and wetland habitat, air quality, recreational river fishing, and recreational river rafting. Together, the estimated avoided costs and environmental benefits comprise a more complete representation of the value of efficiency than water agencies often use, by comparing efficiency to other water supply–enhancing options.

The methodologies employed require significant simplifications of the water systems under evaluation and the use of uncertain estimates of the causal effects of water-use reduction and environmental and recreational benefits. To accommodate the significant uncertainties that result, we opt not to develop a single "best-guess" or likeliest estimate of the value of efficiency. Instead, we take an exploratory modeling approach and evaluate the models under a wide range of plausible assumptions. The results are ranges of possible efficiency benefits. We demonstrate that, even with a wide range of results, the true value of efficiency is significantly larger than one would estimate if only considering SR avoided costs.

Our principal findings are summarized by Figure S.1, which shows ranges of efficiency valuation results as more benefits are accounted for. Each box-and-stem result in the figure represents the present value (PV) of 1 million gallons of efficiency savings per year. The results are derived from 1,000 runs of the models under a wide range of assumptions.

Figure S.1
Present Value of Short-Run, Long-Run, and Total Avoided Costs; Short-Run, Long-Run, and Total Environmental and Recreational Benefits; and the Sum of the Total Avoided Costs and the Total Environmental and Recreational Benefits

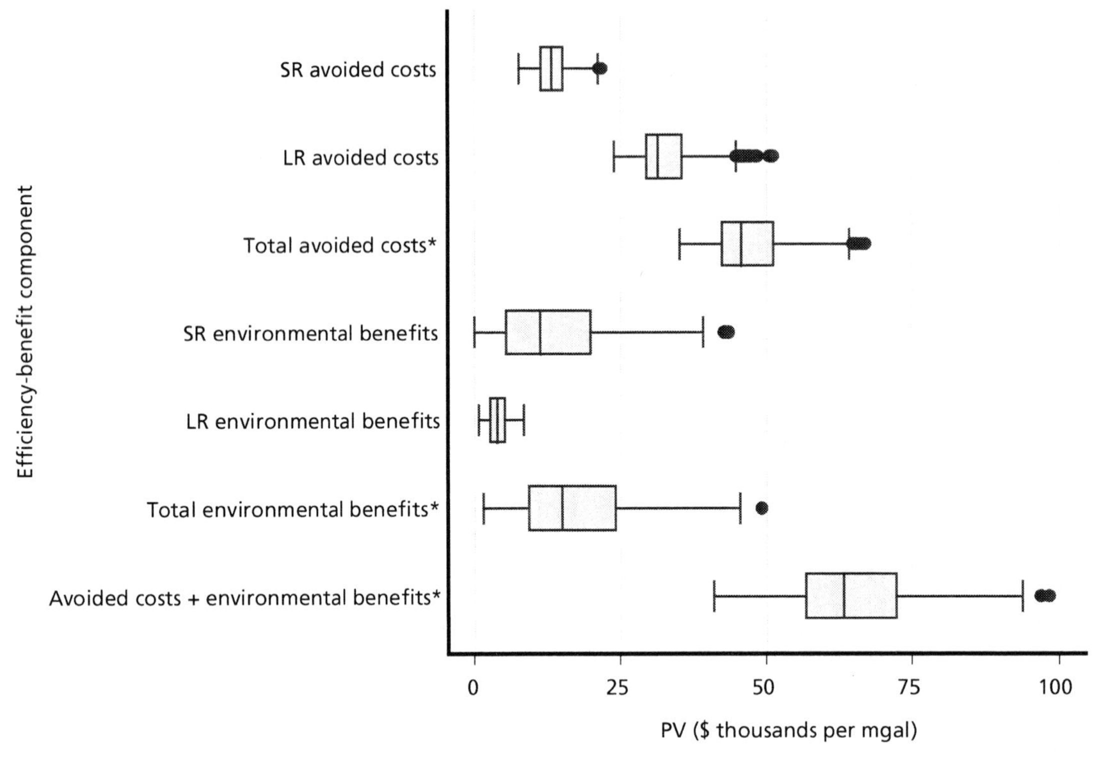

NOTE: Asterisks indicate summation estimates. Each box represents the results of 1,000 model runs and shows the lower quartile (left edge), upper quartile (right edge), and median (inner line) of the results. The dots and stems depict the remaining range of the results.

RAND *TR504-S.1*

For example, considering only the SR avoided costs to Denver Water due to efficiency suggests savings of between $8,000 per mgal and $13,000 per mgal (top box). Adding the LR avoided costs increases the estimate to between $25,000 per mgal and $50,000 per mgal (third box from the top). The value of environmental and recreational benefits adds up to another $50,000 per mgal of benefit (sixth box from the top). After including all avoided costs and environmental and recreational benefits, the estimates suggest a range of marginal benefits between about $41,000 per mgal and just about $100,000 per mgal (bottom box). Although the range of this estimate is quite large, the lower bound of the range ($41,000 per mgal) is about double that of the upper range of the SR avoided costs ($22,000 per mgal). This suggests that, if an agency were to consider only the SR avoided costs, they could be undervaluing efficiency by 50 percent or more. These benefits are even larger if the efficiency induced by a specific program occurs primarily during the summer months—when demand is greatest.

We next compared these efficiency valuations to the economic characteristics of a set of efficiency programs proposed by Denver Water to help meet its 10-year conservation goals. We find that evaluating only the SR avoided costs leads to the conclusion that many water-efficiency projects that are already a part of Denver Water's 10-year conservation plan are not cost-effective. When LR avoided costs and efficiency and recreational benefits are estimated and added to the marginal-benefit calculation, an additional five programs are cost-effective. All but two Denver Water programs were estimated to be cost-effective using this efficiency valuation. Finally, we find that it is critical to consider the timing of projected water savings from efficiency programs. Water savings from programs that concentrate savings during summer months, when water is scarcer, should be valued more highly than savings from programs that lead to more uniform water savings throughout the year.

Acknowledgments

The project team would like to thank the William and Flora Hewlett Foundation for its generous support of this project and the two reviewers for their helpful comments. The project team benefited from helpful discussions with Bart Miller and Taryn Hutchins-Cabibi of Western Resource Advocates and Greg Fisher and Mary Price of Denver Water. The team would also like to recognize the California Urban Water Conservation Council and research teams responsible for the development of the avoided-cost and environmental-benefit models used in this study.

Abbreviations

AC model	avoided-cost model
AF	acre-foot
afy	acre-feet per year
CARs	Computer Assisted Reasoning system
CFS	cubic feet per second
CO	carbon monoxide
CUWCC	California Urban Water Conservation Council
EB model	environmental-benefit model
EEED	Environment, Energy, and Economic Development Program
EIA	Energy Information Administration
EIR	environmental-impact report
ET	evapotranspiration
FERC	Federal Energy Regulatory Commission
IRP	integrated resource plan
ISE	Infrastructure, Safety, and Environment
LR	long run
mgd	million gallons per day
NO_X	nitrous oxide
O&M	operation and maintenance
PM	particulate matter
PV	present value
SO_2	sulfur dioxide
SR	short run

taf	thousand acre-feet
USACE	U.S. Army Corps of Engineers

CHAPTER ONE

Introduction

Increasing water-use efficiency is an important component of prudent water management for regions of growing water demand, such as the western United States. Water efficiency can offset demand growth that would otherwise occur in expanding urban regions, and this moderated demand can reduce the need to develop or acquire new supplies. Urban water demand in the Los Angeles region, for example, has remained largely flat from 1985 to 2005, despite an increase of 5 million people (Metropolitan Water District of Southern California, 2005). The large decrease in per capita water demand resulted in part from ongoing replacement of older, water-using devices with newer, efficient options (passive conservation). Some improvement, however, was the direct result of agency water-efficiency programs (active conservation or efficiency).

It is widely recognized that significant, active conservation potential exists throughout the West. Gleick et al. (2003), for example, provided a comprehensive assessment of opportunities in California and concluded that cost-effective efficiency in the urban sector could reduce demand by 35 percent. Another recent study commissioned by the CALFED Bay-Delta Program, suggests potential water use–efficiency savings of up to 25 percent (CALFED, 2006). Many other agencies in the West are laying plans to improve efficiency in their service areas. Water efficiency and conservation, for example, will play central roles in Denver Water's long-term plan to ensure reliable water supplies through 2050 (Denver Water, 2004; Denver Water, 2006b).

Denver Water recently projected that the population in its service area will increase by more than 50 percent by 2050. Under these projections, the utility expects average annual water demand to exceed average annual supply by 2016 (Denver Water, 2002). To address this projected shortfall, the utility plans to build two new water-supply projects and expand its water-efficiency and conservation programs. As in California, efficiency in Denver will clearly play a role in ensuring future water-supply reliability.

This study addressed an important water-management question: How much water efficiency should a water agency promote in its service area? To answer this question, one must understand how much an efficiency program will cost to implement, how much water it will save, and the value of that water savings to the agency and society. This information can then assist water agencies in comparing efficiency potential to other water-management options to arrive at an appropriate portfolio of actions, given the goals and constraints of the region. This report describes and applies a comprehensive economic approach to valuing water savings. We demonstrate how this approach can provide better guidance to water agencies seeking to exploit water-efficiency potential.

Common Approach to Valuing Efficiency Programs

Water agencies typically evaluate a wide variety of management strategies to ensure sufficient future supply when preparing their long-term water plans, including supply-augmentation options, groundwater–surface water conjunctive use, recycling, and demand management or water-use efficiency. Many criteria are used when weighing the merits of different water-management strategies, although standard planning practice suggests that all else being equal, least-cost resources or strategies to meet a particular objective should be developed first. In the case of ensuring sufficient water supply to meet demand, this approach leads agencies to pursue efficiency programs when they are perceived to be less expensive than other supply-enhancing options. Stated another way, water agencies seek to develop efficiency programs that cost less, on an annualized basis, than the next available water-supply option.

This approach to prioritizing agency investments can be problematic because it implies that the complete value of improved efficiency is captured by the marginal cost of the next least-expensive option for increasing water supplies. Efficiency, however, not only replaces the need to develop or acquire new supply, it also reduces the amount of water that must be transported to the end user and disposed of. Such additional avoided costs are not reflected in a simple cost comparison of the cost of efficiency to the cost of an alternative supply. Second, efficiency can also change the trajectory of future demand and needed new supply, and, if efficiency slows demand growth, expensive new supply projects may be deferred or downsized. These savings to a utility can be sizable and also would not be reflected by the simple comparison to the cost of the next available water-supply project.

There are other benefits of efficiency that accrue to society, such as those to the environment or recreation. These benefits ought to be considered when evaluating the net cost of an efficiency program. The greater these nontangible benefits are, the more important it is to include them in evaluations of the merits of efficiency programs.

Systematic Assessment of Efficiency Benefits

In this report, we describe and demonstrate a more comprehensive and systematic approach to assessing the benefits of an efficiency program. This approach is based on economic cost-benefit analysis that compares the cost of efficiency programs to its benefits to the water agency, its customers, and greater society. In this context, the benefits include the avoided costs of provisioning water to end users and the environmental and recreational benefits of not extracting water from surface rivers and streams.

Avoided Water Costs

Avoided water-provisioning costs are those that would be borne by a utility if demand were not reduced by improved efficiency. Following microeconomic theory, these costs can be disaggregated by time period (Young, 2005). In the short run (SR), the number and capacity of water-system facilities are fixed. SR avoided costs are those due to reductions in the operating costs of these facilities due to demand reductions. In the long run (LR), new facilities can be built or existing facilities can be increased in size. Thus, the LR avoided costs include changes in the capital and operating costs of future supply investments. The combination of the SR and LR costs represents the total avoided costs to the utility.

Environmental and Recreational Benefits

Water efficiency can lead to benefits aside from avoided costs. The accounting of environmental benefits can also be disaggregated into SR and LR benefits. SR benefits are those due to reduced withdrawals of water from a river or stream using existing facilities. LR environmental benefits are those associated with the delay or downsizing of new supply projects and their associated environmental impacts (analogous to LR avoided costs).

SR environmental benefits derive from lower water extractions from rivers and streams and reduced impacts on water-based environmental and recreational services. Important environmental services include river and reservoir recreation, wildlife habitat in riparian zones and wetlands, fisheries supported by rivers and streams, and good water and air quality. Although it is impossible to identify and value all environmental benefits, many benefits can be monetized using nonmarket economic-valuation techniques such as the hedonic-price, market-observation, and travel-cost methods (Platt, 2001; Young, 2005).

Riparian habitat and wetlands are valuable to society because they are ecologically productive and support various plant types and aquatic species. Moreover, riparian habitat and wetlands contribute to water quality, flood control, recreation, and wildlife habitats (National Resource Council, 2005). Riparian zones require sufficient water to sustain their healthy ecosystem function. Reductions in river flows lead to smaller riparian zones and reduced habitat. The hedonic-price method can then be used to approximate a value of these lost environmental benefits by equating the market price for a conservation easement for pristine riparian habitat to the value of the service provided by the unimpaired riparian habitat (Platt, 2001).

Healthy river and lake fisheries provide both sustenance and economic value to society. Such benefits are strongly related to lake levels and natural flow rates, and water extractions for municipal and agricultural water use may diminish these benefits. A portion of the value of fisheries can be estimated using the travel-cost method, which equates the economic benefits of healthy fish populations to the estimated expenditures by persons traveling to the rivers and lakes (Platt, 2001).

Air quality and the impacts of water system–related energy use on it can be important to a region's public health and economic activity. As most water utilities use electric motors to pump some of their water supply to customers and producing the electricity used in these motors results in several types of air pollution, demand reductions also lead to air-quality improvements. In regions of the country with significant air-quality concerns, such as California, there is significant value in reducing emissions of pollutants. The economic impacts of some pollutants, such as nitrous oxides (NO_X, a precursor to smog), is set through tradable markets for emission permits. Other pollutants that contribute to climate change, such as carbon dioxide, do not yet have market values but are likely to soon in many jurisdictions. A lower-bound marginal impact of reducing pollutants through water efficiency can thus be estimated by observing existing markets for pollution permits.

Last, recreational services, such as river rafting and boating, are valuable not only to the users but also to the local and regional economy through the economic activity associated with their use (e.g., equipment purchases and other trip expenses). The quality and economic benefit of water-based recreation can depend on water availability that affects river and stream flows and reservoir or lake levels. Water conservation can thus lead to greater water availability and thus greater recreational services and economic benefit (Coughlin et al., 2006). As it can with fishing, the travel-cost method can be used to evaluate these benefits.

Demonstration of Comprehensive Approach

In this report, we explore the use of new tools in conjunction with an economic framework to evaluate the benefits of water efficiency in the western United States. To demonstrate how water agencies throughout the intermountain West can use this approach to value efficiency and determine how much efficiency to promote through agency programs, we performed a case study of the Denver Water utility service area. Denver Water was not a formal partner in this study, and utilities opting to adopt this approach will likely have access to better data than we had.

This methodology relies principally on two models developed for the California Urban Water Conservation Council (CUWCC). These models were designed to assist water-management agencies in developing and implementing cost-effective water-efficiency management best practices in California. As we show below, the tools can be adapted to apply in non-California settings such as Denver.

The first model used, the AC model, provides a generic analytic framework for calculating the SR and LR avoided costs that a utility would face as a result of efficiency investment within its service area (A&N Technical Services Inc. and Gary Fiske and Associates, 2006). The second, the EB model, was developed specifically for California, and it estimates some environmental benefits that result from efficiency in California through previously estimated values for the environmental and recreational impacts of water use and their respective valuations (Coughlin et al., 2006). Used together, these models can provide a more comprehensive estimate of the benefit of water-use efficiency. This information, coupled with agency efficiency-program cost estimates, can provide guidance for program development and implementation choices.

Due to the inherent uncertainty in estimating future efficiency benefits and lack of all data needed, we exercise the two models using an exploratory modeling approach (Bankes, 1993). Exploratory modeling addresses uncertainty through the evaluation of large numbers of scenarios or cases. At the end, we compare the standard output of these two models with utility-specific projections on efficiency-program costs and savings to suggest which efficiency programs are cost effective under various assumptions about future conditions.

Report Organization

In Chapter Two, we provide background and necessary detail about our case study region: Denver Water. Chapter Three describes the methodology and the details of how we used the CUWCC models to evaluate efficiency benefits in Colorado. Chapter Four presents our finding, and Chapter Five provides some concluding remarks. Appendixes A and B describe the technical details of the avoided-cost and environmental-benefit modeling. Appendix C provides a brief technical discussion of the impacts of alternative supply and demand projections on calculations of LR avoided costs. Appendix D provides more detail on Denver Water efficiency-program costs.

CHAPTER TWO

Denver Water Case Study

Introduction

To demonstrate how an economic framework can be used to evaluate water use–efficiency benefits across the intermountain West region, we selected Denver Water, the main utility serving the Denver metropolitan region, as a case study. We used publicly available data for this demonstration. Use of data typically available to a utility, but not released publicly, would improve the accuracy of such an analysis.

This chapter begins by describing projected water demands and supplies in the Denver region through 2050. Next, it describes each of the main sources of water. It ends by describing the environmental and recreational benefits of these water sources.

Supply and Demand Projections

Denver Water provides water to residents of the city and county of Denver and has additional water-supply contracts with several Denver suburbs. In sum, Denver Water serves more than 1 million people, which is almost a quarter of Colorado's population (Denver Water, 2002; U.S. Census Bureau, 2007).

Like many water utilities in the western United States, Denver Water faces a growing population but limited, and possibly declining, water supplies. By 2050 (Denver Water's long-term planning horizon), Denver Water expects population in its service area to increase by more than 50 percent (Denver Water, 2002). When combined with population growth in the surrounding areas that have supply contracts with Denver Water, the total population growth approaches 60 percent by 2050.

Denver Water is currently undergoing several supply-expansion projects, including construction of a recycling plant and gravel water-storage pits. With these projects, it anticipates that its supply will level off at 375,000 acre-feet (AF) per year (afy) (or 335 million gallons per day [mgd]) by 2015 (Denver Water, 2002, 2004). Using their current projections for population and water use (including conservation due to the natural replacement of water-using devices), Denver Water estimates that water demand will exceed supply in 2016 and the supply deficit will continue to grow. By 2030, they project a 34,000 afy or 30.4 mgd shortfall, and, at the end of its planning horizon in 2050, the deficit is 78,000 afy (69.6 mgd) (Denver Water, 2004) (Figure 2.1).

Figure 2.1
Projected Total System Demand and Supply Without New Supply Projects

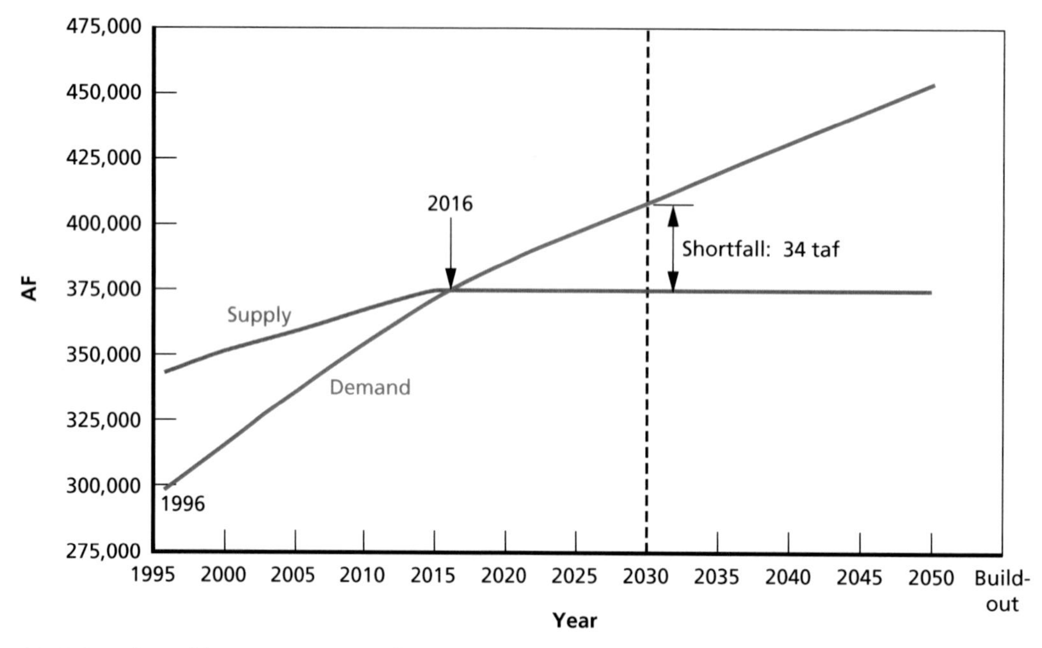

SOURCE: Adapted from Denver Water (2004).
NOTE: taf = thousand acre-feet.
RAND *TR504-2.1*

Denver Water's integrated resource plan (IRP) process, which concluded in 2002, sought to address these concerns with a plan to balance future system demand and supply using a mix of efficiency programs and new water-supply projects. Denver Water is currently finalizing a 10-year conservation plan that intends to reduce per capita consumption by 22 percent and meet the conservation savings specified in the IRP (Denver Water, 2006b). To increase supply, Denver Water has initiated the planning and permitting process for a new supply project in its Moffat system, called the Moffat Collection System Project (hereafter the Moffat Project). This project will develop the additional 18,000 afy (16.1 mgd) of supply needed by 2030 and also relieve operational constraints in the Moffat system (described below). Denver Water also seeks to improve supply reliability in the coming years through the development of more conservation than would be necessary to supply new customers.[1]

Both of these strategies will require significant investment. Denver Water expects to spend more than $240 million on the Moffat Project over the next 10 years (Denver Water, 2006a). According to its plans, this expenditure will still not cover the full construction costs, which it has not yet tallied. Based on a recently completed supply project of similar size in Colorado, the full costs of the Moffat Project could range from $300 million to $450 million (Bureau of Reclamation and Colorado Springs Utilities, 2004). Denver Water projects spending $155 million on its conservation plan over the next 10 years. In addition, consumers may spend up to

[1] A current concern of Denver Water and other water agencies is the effect that efficiency improvements would have on the ability for water users to conserve water further during drought periods. So-called *demand hardening* varies by region and is currently the subject of a Denver Water study.

$255 million on mandatory and voluntary measures, such as changing landscapes and purchasing water-efficient appliances (Denver Water, 2006b).

In addition to the near-term Moffat Project, Denver Water's IRP calls for an additional 31,000 afy (27.7 mgd) of new water supply in its long-range planning horizon (2030–2050) (Denver Water, 2002). Denver Water has not identified any specific projects for this water supply yet (we call this the "second project" in this analysis), but the IRP discusses the following long-term supply options: increased system storage, new stream diversions, and effluent reuse.[2] An important note is that the first two options are potentially costly and controversial. Any water development on the western slope, which includes all the stream diversions considered except the South Platte, would potentially involve construction of a new tunnel, may require construction in wilderness areas, and may pose difficulties for threatened and endangered species in these rivers (Denver Water, 2002).

Sources of Supply

Denver Water obtains water from three major systems: South Platte, Roberts Tunnel, and Moffat (Figure 2.2). In addition to providing water supplies for Denver Water and other users, the natural flows within each of these systems support important ecosystems and recreational settings for visitors. Riparian habitats and wetlands, for example, cover a very small percentage of the land area in Colorado, but they directly or indirectly support more than half of the wildlife species in the state (Colorado Division of Wildlife, 2007; Culver, 2001). Colorado rivers and streams also support various types of recreational services ranging from rafting to fishing to camping. Fishing, for example, adds more than $1.5 billion to Colorado's economy per year (BBC Research and Consulting and Colorado Division of Wildlife, 2004).

The South Platte system, the largest in the Denver Water system, collects snowmelt from the upper South Platte River basin and stores the water in three large reservoirs (Denver Water, 2005). After water is released from these reservoirs, it flows via the South Platte River and feeds two of Denver Water's treatment plants: the Foothills and Marston plants. The treated water then enters the distribution system, where it is delivered to Denver Water customers. Wastewater is then treated and released back into the South Platte River. Note that this wastewater cannot be reused by Denver Water.

During summer months, the South Platte River supports scenic and attractive runs for rafters, canoeists, and kayakers. It also is habitat to nationally renowned cutthroat, brown, and rainbow trout populations (Christopherson, undated[b]). Some segments of this river are identified by the Colorado Division of Wildlife as an important quality trout fishery in the state, attracting people from all over the region. In most parts, the river supports self-reproducing rainbow and brown trout populations and is designated as Colorado Wild Trout Water (U.S. Forest Service, 2004). Finally, the South Platte River reaches include significant riparian habitat and wetlands.

---

[2] Denver Water has already explored cooperative actions with other water providers in the region to expand current system storage. These proposals could become part of a long-term supply project. Denver Water maintains water rights on the Eagle River and Williams Fork River. In addition to these, the IRP suggests other potential stream diversion projects on the Colorado, Blue, and South Platte Rivers. Finally, Denver Water recently built a water-recycling plant, and expanding this plant or an additional plant could become part of a future supply project.

Figure 2.2
Simple Schematic of Denver Water Sources of Supply and Treatment Facilities

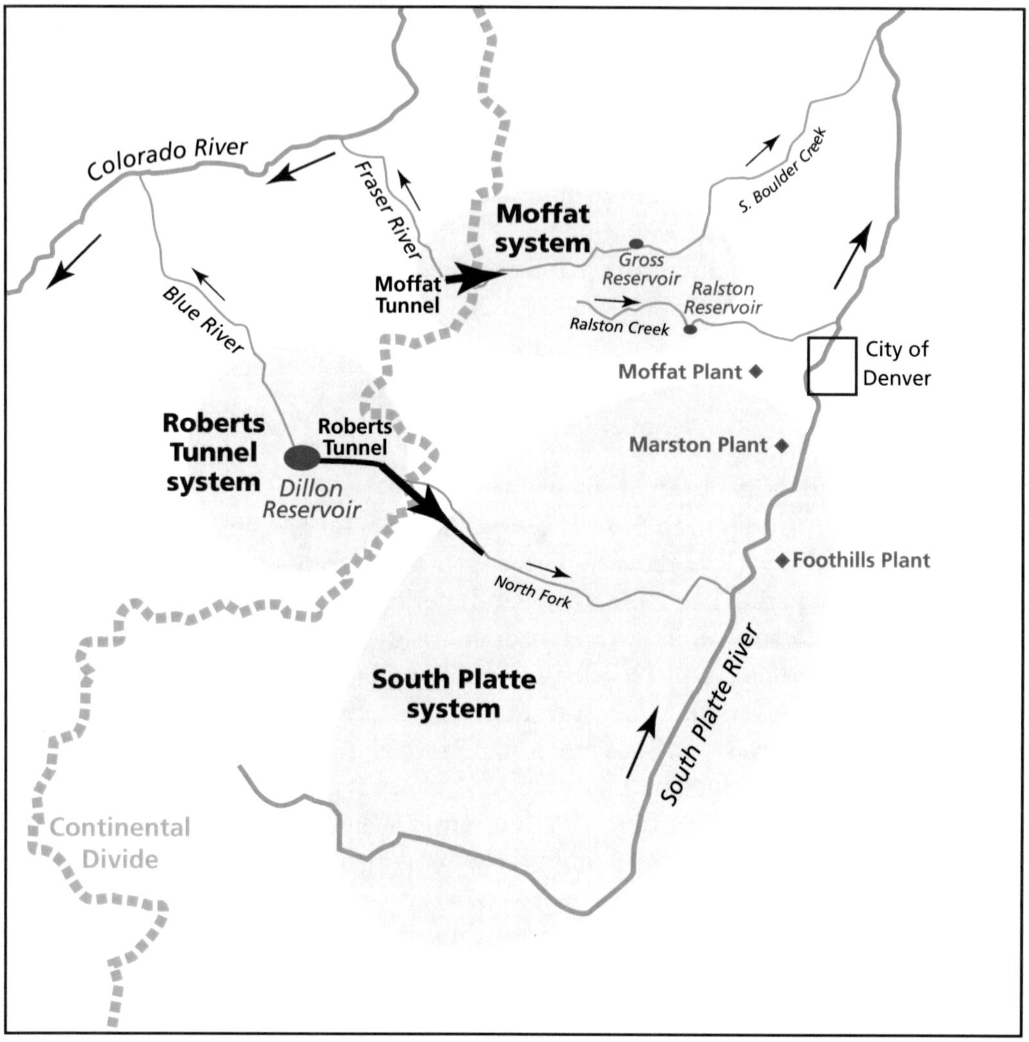

NOTE: Arrows show direction of water flow.
RAND *TR504-2.2*

The Roberts Tunnel system is the second-largest system and is located on the west slope of the Continental Divide (Denver Water, 2005). The Blue River is the main raw-water source in the Roberts Tunnel Collection System. This river originates in the mountains above Dillon, Colorado, and flows about 40 miles to its confluence with the Colorado River. The collection system stores snowmelt from the upper Blue River watershed in Dillon Reservoir. Water is pumped through the Roberts Tunnel and fed into the South Platte River, where it also supplies the Foothills and Marston treatment plants.

The Blue River supports substantial river-based recreation. Dillon Reservoir, the largest water-storage facility in the Denver Water system with a surface area of 3,200 acres and a shoreline of 27 miles, supports scenery, boating, canoeing, fishing, and camping (Denver Water, undated[a]). The Blue River is also habitat to brown and rainbow trout. Several sections of this river are fishable, and some of it, including the section below the Dillon Reservoir, is designated as a Gold Medal fishery by the state (Cutthroat Anglers, undated).

The Moffat system is the smallest of the three Denver Water systems and is operated separately from the two others. Water in the Moffat system originates on the west slope of the Continental Divide in the upper Fraser River basin. The Moffat system collects raw water from several tributaries of the Fraser River and pumps it through the Continental Divide via the Moffat Tunnel. This water is then stored in two smaller reservoirs (Gross and Ralston reservoirs) and eventually supplies the Moffat treatment plant using South Boulder Creek and Ralston Creek for distribution (Denver Water, 2005). The Moffat treatment plant is not connected to the other collection systems. When supplies are low in times of prolonged drought, Denver Water sometimes needs to shut down the Moffat treatment plant to maintain sufficient water levels in their reservoirs to ensure adequate water supply in future years (Denver Water, 2004). This operational constraint is an important consideration in pursuing a new supply project in this portion of the system.

The Fraser River supports both fishing and ice fishing. It is habitat to Cutthroat, Brown and Rainbow trout populations (Christopherson, undated[a]). Many communities depend on this river to attract visitors during the summer months (American Rivers, undated).

CHAPTER THREE

Case-Study Methodology

Introduction

This study evaluates a more comprehensive economic approach to valuing water use–efficiency program benefits than is often used by water utilities. This method explicitly estimates avoided utility costs and some of the avoided environmental and recreational impacts associated with demand reduction induced by efficiency programs. When estimated properly, these benefits are additive both on an annual basis and when discounted and summed over the planning horizon. The total value of water efficiency is the sum of the avoided costs and the environmental and recreational benefits:

$$\text{Total benefit} = \left(\text{SR avoided costs} + \text{LR avoided costs}\right) \\ + \left(\text{SR environmental benefits} + \text{LR environmental benefits}\right).$$

Estimating Avoided Costs

We used the CUWCC AC model to estimate the annual SR and LR avoided costs due to efficiency in Denver Water's service area from 2007 to 2050. SR avoided costs are those due to reductions in existing water-system operating costs due to demand reductions. LR avoided costs are those due to the deferral or downsizing of future supply projects. In the following sections, we describe the basic methodology for calculating SR avoided costs, LR avoided costs, and environmental and recreational benefits.

Short-Run Avoided Costs

The AC model estimates SR avoided costs by calculating the expected savings of operating expenditures due to a reduction in water demand. For a simple system with only one set of facilities required to provide water supply and treat the wastewater, the avoided cost would be equal to the sum of the marginal cost of operating each facility. In more complex systems, different facilities provide the marginal supply (e.g., the last unit of water) at different times of the year and under different hydrologic conditions. In such cases, the avoided cost would be the sum of the marginal costs for each facility weighted by the probability of time that they provide the marginal supply (or their "on-margin" probabilities). The AC model thus requires the user to estimate on-margin probabilities for each of the system components under several hydrologic conditions. It then combines this information with marginal operating costs to calculate expected savings across the entire system.[1]

Long-Run Avoided Costs

LR avoided costs are those associated with the deferral or downsizing of future supply projects due to marginal demand reductions. The AC model takes as inputs the costs and schedule of new supply projects and specification of whether each project can be downsized or deferred.[2] The AC model then calculates the value to the utility of such changes due to demand reductions.

For deferrals, the model utilizes a user-supplied peak-demand forecast to calculate the deferral period as

$$\text{deferral period}[\text{time}] = \frac{\text{demand reduction at planned new supply online date}[\text{flow}]}{\text{change in peak demand}[\text{flow}/\text{time}]}.$$

Note that only demand reduction during peak periods leads to LR avoided costs. This is important, as some efficiency programs, such as those inducing outdoor irrigation efficiency, lead to greater savings during peak periods.

This calculation assumes that average system supply does not decline over time. To evaluate the sensitivity of avoided-cost estimates to alternative demand forecasts and supply projections, we modify the AC model to allow system supplies to decline over time to reflect possible reductions in available supply (due to climate change, changes in minimum stream-flow requirements, or changes in water rights, for example).

After determining the deferral period, the model uses the project cost information to calculate the annualized cost[3] of each future supply project before and after any deferral. This distributes the initial construction cost over the lifetime of the project. The difference in the annualized costs is then the LR avoided costs in each year of the project lifetime. The magnitude of this difference depends both on the deferral period and on the discount rate as provided by the user. For projects that can be downsized, the user chooses the amount by which it can be downsized, and the model then uses this information to calculate a new annualized cost by reducing capital costs proportionately.

The AC model produces annual estimates of SR and LR avoided costs during the peak and off-peak seasons for each year in the planning horizon. The model calculates these values in both nominal and real terms. To evaluate the effect of these uncertainties on the overall value of water-use efficiency, we calculated the present value (PV) of all future avoided costs.

---

[1] A more accurate but more technically complicated approach to this calculation would be to use a simulation model to project which system components provide the marginal unit of water during each time period. The SR avoided cost due to efficiency would then be the cost of operating these facilities to supply a unit of water during that specific period and hydrologic condition.

[2] In general, if system demand is still expected to exceed supply in the future after completing the project, the project is perhaps best considered deferrable. If, at the end of the planning horizon, some portion of the project is not required, the project can be downsized.

[3] Annualized costs convert a sum of money in PV to an equivalent future stream of payments over a specified period and interest rate.

Representing Denver Water in the Avoided-Cost Model

For the purposes of this analysis, we represented Denver Water's system as three separate water sources and distribution paths that correspond to the collection systems described in Chapter Two (South Platte, Roberts Tunnel, and Moffat) as well as three treatment plants (Foothills, Marston, and Moffat). Denver Water's publicly available documentation has limited information on system operation; thus, we were unable to use a more complicated representation that would include elements such as the dispatch order of individual reservoirs. While our analysis used a considerable simplification of the system, it still followed the general classifications as described in Denver Water documentation. The model does differentiate between the on-margin probabilities and operation costs for the three different collection systems, and Denver Water's annual budget report provides more detailed information on operating costs across the three collection systems and treatment plants (Denver Water, 2006a). For complete details on how this analysis represents the Denver Water system, see Appendix A.

We used Denver Water's plans articulated in the IRP as the baseline projection for supply and demand. With this information, we constructed simple supply and demand forecasts using the projected levels of demand, initial supply, new supply, and conservation. We treated Denver Water's new supply plans as two separate projects. The first to come online (the Moffat Project) will provide 18,000 afy (16.1 mgd). The second to come online (the second project) will provide 31,000 afy (27.7 mgd). We estimated the costs of new future supply based on a recently completed, comparable project in the Rocky Mountain region (Bureau of Reclamation and Colorado Springs Utilities, 2004).

We also allowed several demand and conservation parameters to vary, and we estimated the online dates for each supply project. These assumptions closely follow the plans described in Denver Water's major planning documents. Figure 3.1 shows the projected demand (bars) and supply (line) from 2007 to 2050 under base-case conditions as projected by Denver Water (Denver Water, 2004). The jumps in supply at 2015 and 2030 are due to new water from the Moffat Project and Second Project, respectively.

Changes in either the demand or the supply will lead to changes in the scheduled supply increase from the two projects. To illustrate how this methodology reflects such changes, Figure 3.2 shows the original schedule of demand and supply (light gray bars and dashed line) and alternative schedules in which demand in 2050 is reduced by 5 percent. As a result of the reduced demand projection in this example, the Moffat Project does not come online until 2023 and the second project does not come online until 2042. If demand were to occur faster than anticipated, the projects would be needed sooner. Exogenous changes in supply (due to hydrologic conditions or legal developments, for example) would also affect when projects would be required. Appendix C provides a more technical discussion of these effects.

Estimating Environmental and Recreational Benefits

We used the analytic framework of the CUWCC EB model to develop a range of estimates of the economic value of the environmental and recreational benefits that result from reducing withdrawals from Denver Water sources (rivers, streams, and reservoirs).[4] The EB model uses

[4] As this methodology does not measure all environmental and recreational benefits, the actual (and unmeasurable) environmental and recreational benefits may be greater than the range estimated.

Figure 3.1
Supply and Demand Under Nominal Assumptions

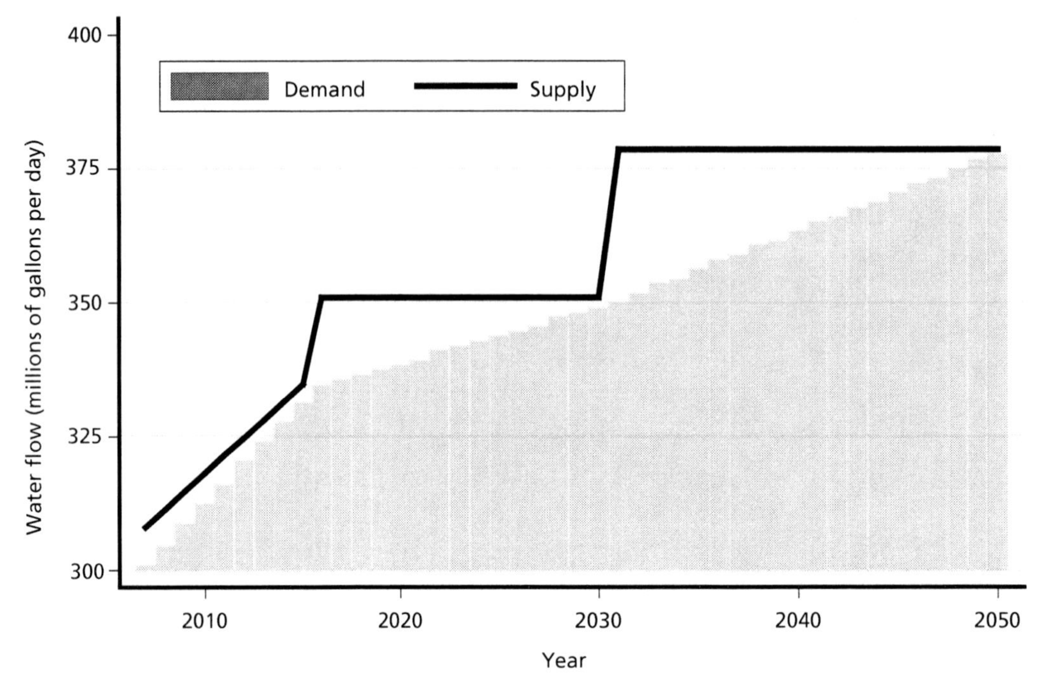

NOTE: The rapid increase in supply between 2015 and 2020 reflects new supply from the Moffat Project. The increase around 2030 is from the second supply project.
RAND *TR504-3.1*

a simple methodology to estimate SR benefits for a wide, but not exhaustive, set of environmental services and recreational activities in California. For each service, the study estimated the environmental impact of water use on the service and then the monetary value associated with change in the service.

The SR benefit due to efficiency is then estimated as the product of the environmental or recreational impact and the value of the environmental or recreational service (Coughlin et al., 2006):

$$\text{Benefit[value/time]} = \text{marginal impact of use[service/flow]} \times \text{service value[value/service]} \times \text{flow reduction[flow/time]}.$$

The EB model produces annual estimates of environmental and recreational benefits during the peak and off-peak seasons for each year in the planning horizon.

The EB model for California addresses environmental and recreational SR benefits from the following services:

- recreation at reservoirs
- habitat in riparian areas and wetlands
- fisheries supported by river flows
- habitat supported by suitable salinity levels in the San Francisco Bay-Delta
- air quality (affected by emissions of air pollutants).

Figure 3.2
Supply and Demand Under Nominal Assumptions

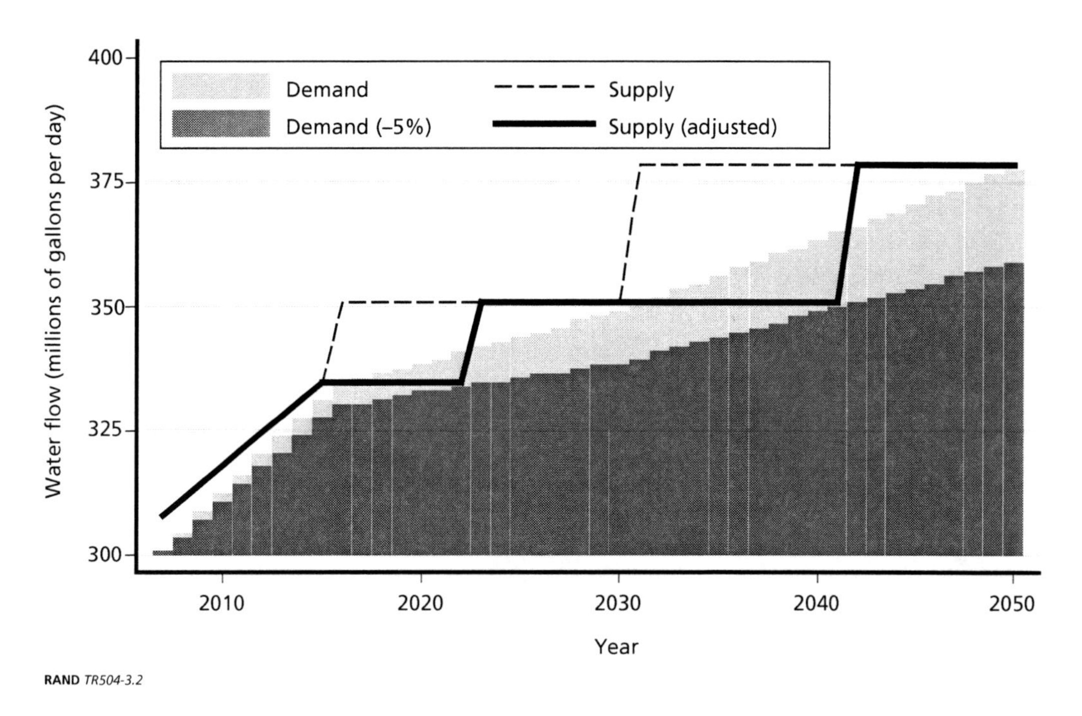

RAND *TR504-3.2*

For our Denver Water case study, we consider riparian and wetlands habitat, fish habitat, air-quality impacts from energy use in water distribution and treatment, and nonangling river recreation (services) as sources of recreational benefits. Table 3.1 lists each service and valuation method, impact metric, and valuation metric used. These are explained in the text following the table, and Appendix B provides more detail on each valuation-estimation procedure.

Riparian Zones and Wetlands

To estimate the impact of water use on riparian zones and wetlands in Denver Water's source regions, we followed the general methodology used in the environmental-benefit report

Table 3.1
Environmental Service, Valuation Methodology, Impact Metric, and Valuation Metric Considered in the Case Study

Environmental Service	Valuation Methodology	Impact Metric (service/flow)	Valuation Metric (value/service)
Riparian and wetlands	Property value as proxy for ecosystem value	Acres/mgd	$/acre (annualized)
Trout fisheries (recreational benefit of angling)	Economic impact attribution	$/mgd [value/flow][a]	
Air-quality impacts from water pumping	Market analysis	Tons of NO_x/mgd	$/tons of NO_x
River recreation (nonangling)	Modified travel-cost method	User-days/mgd	$/user-day

[a] Note that, for trout fisheries, we do not estimate both the service per flow and value per service; instead, we estimate the value of the flow directly.

(Coughlin et al., 2006). Water availability's impact on the services provided by riparian habitats and wetlands is based on assumed water needs per acre of riparian habitats and wetlands. The EB model uses annualized per-acre purchase prices for different habitat- and wetland-acquisition projects as a very rough proxy for the dollar value of the services provided by riparian habitats and wetlands.[5]

In our application, we first estimated the buffer area of the rivers, including riparian habitat and wetlands, by multiplying the length of the rivers assumed to be impacted by water-use reduction by Denver Water by an assumed average buffer-zone width. In the uncertainty analysis, we allowed average widths of between 25 and 75 feet, with a base-case estimate of 50 feet (the value assumed for the California EB analysis).[6]

Next, we estimate the consumptive water needs of riparian habitat. We made adjustments to these water needs to reflect monthly variation in temperature and humidity. We multiply these consumptive needs by estimated attribution factors, to reflect how much of a unit of water used from a river decreases the water availability to the riparian or wetland zone. Finally, we multiplied the attribution factor by the number of acres sustained through a unit of water to estimate the amount of riparian or wetland area impacted by a unit of water use.

The environmental-benefit study used per-acre purchase prices for different habitat-acquisition projects as proxies for their environmental value. Due to lack of data on Colorado riparian or wetland valuations, we considered a wide range of riparian habitat valuations bounded by the annualized average purchase price for California riparian habitat ($215 per acre) and a higher value ($1,000 per acre) to reflect the inclusion of wetlands which are priced in California at about 10 times that of riparian habitat ($2,500 per acre). Table 3.2 presents the impact of water use for each of the three major rivers considered in this case study.

Trout Fisheries

We estimate the economic impact of water conservation on river trout fishing using a methodology similar to that used by the environmental-benefit report to estimate the economic benefits of the anadromous (or spawning) fish habitat in California. The environmental-benefit study multiplied an estimated relationship between fish populations and annual river

Table 3.2
Riparian and Wetland Impact and Valuations, by Water System (for selected months)

River	Month	Impact (acres/mgd)
South Platte River (South Platte system)	Minimum (December)	0.119
	Maximum (June)	0.538
Blue River (Roberts Tunnel system)	Minimum (December)	0.166
	Maximum (June)	0.755
Fraser River (Moffat system)	Minimum (December)	1.147
	Maximum (June)	5.206

[5] Purchase-price data are annualized assuming an amortization at 6 percent interest over 30 years.

[6] Unfortunately, we were not able to obtain more location-specific information for calibrating this and other parts of the environmental-benefit estimation. For this reason, we assume relatively wide ranges for these parameter values. The lack of site-specific information does not detract from the demonstration of the model's applicability, however.

flow and an estimated economic valuation of fish to anglers. The EB model estimates the dollar value of services provided by fish habitat in terms of price per fish caught per year using two approaches: (1) survey reports of the average marginal consumer surplus per fish caught per year and (2) a willingness-to-pay estimate of doubling fish catch for all anglers.

Due to limited data available for Colorado within the scope of this study, we could not directly estimate the environmental impact of water conservation on the environmental service provided by the fish population. Instead, we developed a simple relationship between annual river flow and economic impact (Table 3.3). We use two estimates of the economic impact of recreational fishing in Colorado by the American Sportfishing Association for 2001 and 2003 (American Sportfishing Association, undated[a], undated[b]) and then estimate how much of this change in economic impact is attributable to changes in river flow for each of the three rivers. We next compare the relationships between changes in economic activity between 2001 and 2003 and annual river flows for the same years. Note that the Fraser River annual flow change between 2001 and 2003 is opposite that of the statewide economic impact, so we cannot estimate the impact of use for the Fraser River using our indirect and admittedly ad hoc approach. Appendix B presents the details of this calculation.

Air-Quality Impacts from Water Pumping

The emission calculation in this case study closely follows the methodology used for the California version of the model. When possible, we modify parameters to be appropriate to the Rocky Mountain region. We calculate NO_X emissions associated with water use by multiplying the energy demand for pumping water by the emissions released per unit of energy and the cost of reducing NO_X emissions, the latter used as a rough proxy for the value of emission reductions. In this illustrative calculation, we do not consider other air pollutants that might be relevant to the Denver area, such as carbon monoxide and particulate matter.[7]

We calculated NO_X emission intensity using projections from the Energy Information Administration's (EIA) *Annual Energy Outlook 2007* (EIA, 2007, Table 73). We relied on model default values for the energy-demand rate because we were unable to obtain the needed information from the utility. For the cost of NO_X emissions, we assumed a range of values

Table 3.3
Differences in River Flow and Fishing Economic Impact Between 2001 and 2003 and the Associated Estimated Impact of Water Use Under Base-Case Assumptions

River	Annual River-Flow Difference (mgal)	Difference in Economic Impact ($ thousands)	Impact of Use ($/mgal)
South Platte	2,600	4,200	1,600
Blue River	12,700	900	70
Fraser River	−600	480	—

SOURCE: Calculations based on American Sportfishing Association (undated[a], undated[b]).

[7] The Denver region currently exceeds the EPA's eight-hour ambient ozone standards and is preparing a plan to reduce emissions. NO_X is an important smog-forming pollutant, and reductions will be necessary to meet the EPA plan. The Denver region is currently in compliance for emissions of other criteria pollutants but achieved this status only recently. For nearly two decades, Denver exceeded EPA standards for carbon monoxide (CO) and particulate matter (PM), and emissions of these pollutants remains closely monitored to ensure compliance (RAQC, undated).

based on two estimates. The default value in the model is the average cost of a NO_X emission permit traded in California ($6,000 per ton of NO_X). We used this as an upper bound, and, for the lower bound, we relied on a recent EPA estimate of the marginal cost of controlling NO_X. In the technical analysis for the Clean Air Interstate Rule (EPA, 2005),[8] the EPA estimated the marginal cost of controlling NO_X at $1,250 per ton (Chappell, 2005). Denver is unlikely to face the same challenges in reducing emissions as California did, and therefore the marginal costs of pollution control for Denver are likely to be smaller than the cost of permits traded in California. The estimate of marginal pollution-control costs in the Clean Air Interstate Rule involved power plants in 28 states, primarily east of the Mississippi River. The many power plants covered under this rule are likely to have lower-cost pollution-control opportunities than would a smaller area, such as Denver. Therefore, although the true cost of NO_X pollution control remains uncertain, these estimates provide a plausible range.

River-Rafting Recreation

To estimate the impacts of reduced water withdrawals on river rafting, we followed a methodology similar to the one used by the environmental-benefit study to estimate lake-recreation impacts in California.[9] We tested for a historical relationship between annual commercial river use and seasonal discharge (corresponding to the rafting season) for the Blue River and the South Platte River. The data we used (Colorado River Outfitters Association, undated) did not include figures for the Fraser River, so we did not include this river in our analysis. As shown in Appendix B, there is a statistical, positive relationship only for the Blue River. We assume that this relationship represents the elasticity of recreational use to changes in river flow and is the impact of water use on rafting (0.083 user-days per mgal). For the economic value of rafting, we calculated a statewide average for economic impact from a user-day of rafting (dollars per user-day). We estimated this value using data collected on total user-days of rafting and total economic impact from rafting for the state of Colorado. These assumptions lead to an estimate of the economic value of a rafting user-day to be $267 per user-day. Note that the California study estimated the value of a lake reservoir user-day to be $37 per day. The economic impact of a unit of water extraction during rafting season (May to September) is the product of the marginal impact on rafting days and the value of a rafting day.

It is important to note that this valuation excludes other river and lake recreational services and their associated values and should be considered a lower bound of the total recreational benefit obtained from water conservation.

Long-Run Environmental Benefits

The EB model does not explicitly evaluate LR environmental benefits of efficiency (e.g., those due to deferred or avoided environmental and recreational impacts of new supply facilities). To evaluate the effect of efficiency on reducing these impacts, we added an estimate of annual environmental impact (in current dollars) to the project's annual operation and maintenance

[8] The Clean Air Interstate Rule limits emissions of sulfur dioxide (SO_2) and NO_X for 28 states in the eastern United States and the District of Columbia. For more information, see EPA (2007).

[9] The impacts on the services provided by lake recreation are based on the elasticity of rates of visitation for recreational activities with respect to changes in reservoir surface area. The value of a reservoir recreational user is based on visitation and economic-expenditure estimates.

(O&M) costs. The difference between the LR avoided-cost estimate with and without this additional cost is the LR environmental benefit due to the deferral of the particular project.

As the environmental impact report (EIR) for the Moffat Project has not yet been released and the second project has not yet been identified or evaluated, we considered a broad range of possible impacts of these projects to illustrate how this effect could be important in the total efficiency valuation calculation (Table 3.4). We assumed a range of $500,000 per year to $5,000,000 per year for the Moffat Project and $1,000,000 per year to $10,000,000 per year for the second project. The selection of these ranges is guided by qualitative concerns revealed during the initial scoping of the Moffat Project, including the proposed diversions on the health and functioning of the Fraser River and on the "water-related, recreation-based" economy of the Fraser River Basin, particularly Grand County (USACE, 2003). As quantitative estimates of the environmental impacts of these new projects become available, these assumed ranges of impacts can be significantly refined.

Addressing Uncertainty

Any projection of the benefits of efficiency programs will be uncertain, as future conditions are ultimately unknowable with complete accuracy, and any evaluation method will necessarily make approximations and simplifications of important processes. For the methodologies used in this study, the model representation of the agency's water system will influence the precision and accuracy of the efficiency valuation estimates. The coarser the representation, the less precise the estimates will be. For any system representation, there are many factors that are uncertain:

- *future hydrologic conditions*: the frequency of future wet, dry, and normal conditions
- *future marginal-supply facilities*: the probability that any given facility will be on margin in the future
- *facility operating costs*: the costs avoided when efficiency reduces facility operation
- *cost of future supply projects*: the costs that are deferred or reduced in response to efficiency
- *responses of environmental and recreational services to water extractions*: how much such services are degraded due to water use by a utility
- *valuations of environmental and recreational impacts*: the financial impact on society due to reductions in environmental and recreational service
- *value of environmental impact of future supply projects*: the total annual financial impact that future supply projects will have on society.

Table 3.4
Range of Hypothesized Net Impacts (evaluated at the date of completion) for the Two Denver Water Supply Projects

Project	Low Value ($/year)	High Value ($/year)
Moffat Project	0.5 million	5 million
Second Denver Water project	1 million	10 million

Given the simplifications inherent in the AC and EB models and lack of all necessary data to represent the Denver Water system and its impact on environmental and recreational services, we determined that we would need to evaluate a wide range of values for these uncertain parameters. Furthermore, because accurate probabilistic information was unavailable, we adopted an exploratory modeling approach and developed a large ensemble of modeling cases, with each case reflecting different plausible combinations of values for key uncertainties in the models (Bankes, 1993; Lempert, Popper, and Bankes, 2003). As a result, the analysis presents not a single result, but instead ranges of results that need to be appropriately interpreted. To facilitate this analysis, we connected the AC and EB models to exploratory modeling software.[10]

Relating Efficiency Valuations to Efficiency-Program Planning

An important objective of this analysis is to connect the quantitative estimates of the value of water-use efficiency to actual agency planning decisions. There are several complications in achieving this objective. The AC and EB models provide estimates of the annual value of water-use reduction due to efficiency from the present through the planning horizon for peak and off-peak periods. The benefits of water-efficiency programs, however, are not always uniform throughout the year, nor are water savings constant in the future. Furthermore, cost estimates of efficiency programs have fixed and variable components and are often averaged over a fixed time period to produce multiyear program-cost estimates.

To compare the benefits of efficiency programs calculated by the methodology presented here with utility estimates of efficiency-program costs, we first calculated the PV of the avoided costs and environmental and recreational benefits due to a unit of demand reduction. Because the estimated benefits differ for peak and off-peak seasons, the PV estimate depends on the efficiency program's saving profile. In the analysis that follows, we assume that the efficiency program saves water uniformly throughout the year, then consider the case in which savings occur only during peak periods. We then estimate the PV marginal program costs, making basic assumptions about the relationship between average and marginal costs. Finally, we compare the marginal benefits to marginal costs. A comparison of these costs to other alternative actions provides rigorous guidance as to whether to implement the efficiency program.

[10] We used the Computer Assisted Reasoning® system (CARs™) available from Evolving Logic to facilitate our exploratory modeling. CARs provides a generic analytic environment to generate and manage ensembles of modeling cases in which each case is defined by a specific set of input parameters and the corresponding output values.

CHAPTER FOUR

Results

This report demonstrates new methods that water utilities can use to account more comprehensively for the benefits of water conservation. Utilities can compare these benefit estimates to independent estimates of the costs of various efficiency programs. Assuming no constraints on budgets, utilities may seek to maximize the net benefits of their efficiency programs by investing in measures until the point at which marginal costs equal marginal benefits. Because estimates of marginal costs and benefits are likely to be uncertain and budgets limited, utilities may choose instead to use this information to gauge by how much the benefits of proposed efficiency measures exceed costs. Utilities should have confidence that those programs in which this gap is large are worth doing and use progressively more caution as the cost estimates approach the benefit estimates.

In this chapter, we present the details of a case study estimating the value of water-use efficiency in the Denver Water service area and comparing these valuations to cost estimates of specific programs. We use modified versions of the CUWCC AC and EB models to estimate SR and LR avoided costs and environmental and recreational benefits to water efficiency. We present the results incrementally to emphasize the importance of considering as many benefits as possible and present ranges of benefits to explicitly reflect the significant uncertainty. At the end of the analysis, we compare our estimates of the benefits of efficiency to independent program cost and saving estimates developed by Denver Water in support of its recent 10-year conservation plan (Denver Water, 2006b).

Short-Run Avoided Costs

The standard output of the AC model is a time series of SR avoided costs (using current and nominal dollar valuations). Figure 4.1 shows these results for the Denver Water system under base-case conditions (1) using undiscounted 2005 dollars and (2) discounting the results by 6 percent per year. Recall that the SR avoided costs reflect reduced operating costs for existing water systems. The undiscounted SR avoided cost from improved efficiency is just less than $700 per mgal of demand reduction due to efficiency. Over time, the discounted benefits decrease to about $300 per mgal by 2020 and less than $100 per mgal by 2040. This stream of discounted future benefits is the basis for comparing a PV of benefits to an up-front investment for improved efficiency over time.

Figure 4.1
Short-Run Avoided Costs over Time (undiscounted and discounted at 6 percent per year through 2050)

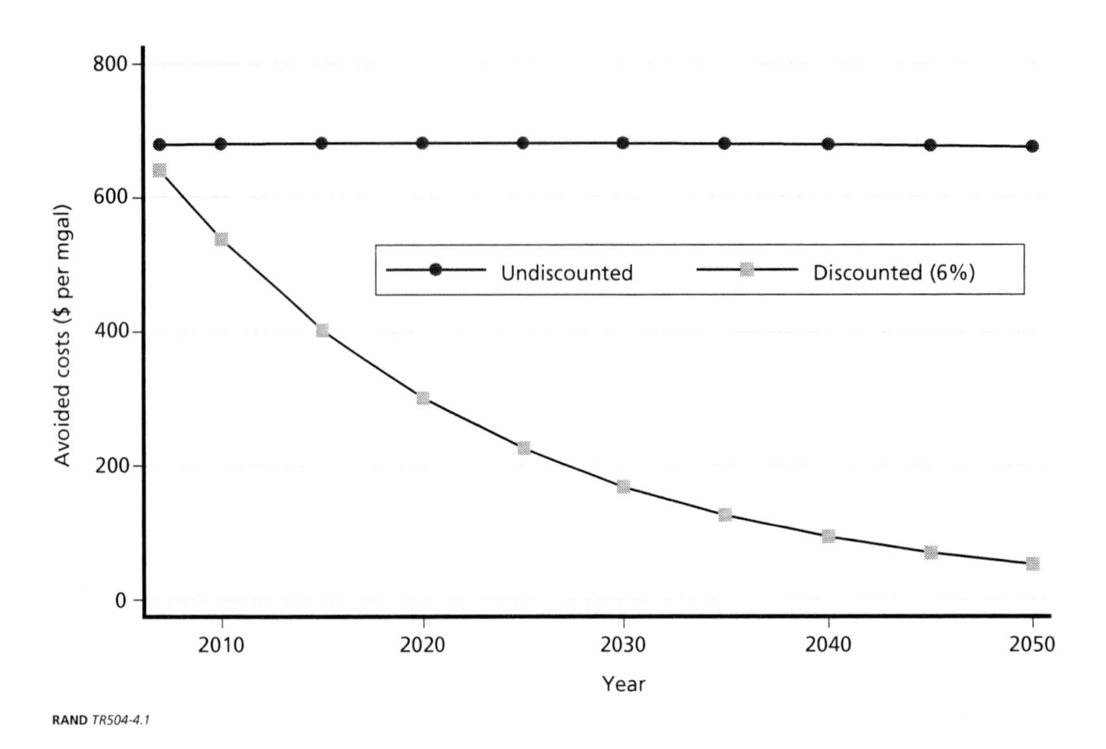

RAND *TR504-4.1*

To compare results across different assumptions underlying this calculation, as well as compare benefits to costs, we sum up the discounted avoided costs by year. Table 4.1 shows the PV of the SR avoided costs for various discount rates in dollars per mgal saved across the years. These results are the PV of a continuous sequence of annual savings of 1 mgal (through 2050). Using a 6 percent discount rate, for example, the PV SR avoided cost is about $10,300 per mgal.

As with all calculations that consider costs and benefits in the future, the results are highly sensitive to the discount rate chosen, and the correct discount rate to use for cost-benefit analyses depends on the specific context (Frederick, Loewenstein, and O'Donoghue, 2002). If the discounting is to represent an agency's cost of capital for use in implementing a program, then a discount rate of 8 percent or more may be appropriate. If the discount rate is to represent

Table 4.1
Present Value of Short-Run Avoided Costs (saved through 2050), by Discount Rate

Discount Rate (%)	PV SR Avoided Costs ($/mgal)
2	19,400
4	13,700
6	10,300
8	8,100

societal preferences for intertemporal consumption, then a discount rate of 3 percent might be more appropriate. Given that this problem represents some of both, we present results using the 6 percent discount rate—also the default value for the application of the AC and EB models in California.

As discussed above, these SR avoided-cost calculations rely on approximations of the Denver Water system and assignment of model parameters that may be difficult to know with certainty. The key factors determining the SR avoided costs include the following:

- current variable costs for the facilities used to provide the water
- changes in the variable costs
- probabilities of types of hydrologic conditions (e.g., wet, normal, dry)
- the probabilities that each system component is the marginal facility by year.

All but the first of these factors are highly uncertain. For example, future costs of water provisioning may increase in response to changing energy prices, regulatory requirements, or unanticipated system repairs. The probability of different hydrologic conditions during the planning horizon affects the amount of water available to the system and is uncertain both due to natural, long-term variability in climate and due to any changes to this variability in response to human-induced climate change. Finally, which facilities are on margin under different hydrologic conditions is uncertain for a variety of reasons, including unforeseen operational conditions, requirements, or procedures and the relative available supply across Denver Water's system.

To consider how this uncertainty affects the SR avoided-cost estimate, we evaluated the AC model numerous times using different values for the key factors shown in Table 4.2. The first two factors control the probabilities of normal, wet, and dry years. The next six factors control the on-margin probability ratio for each of the three systems under normal, wet, and dry years. For example, the nominal values for the normal-year factors specify that the South Platte system is on margin 33 percent of the time and that the Roberts and Moffat systems are each on margin the same amount (50 percent) and thus are also on margin 33 percent of the time (50% × [100% – 33%]). The last factor specifies the annual increase in operating costs for

Table 4.2
Uncertain Factors Affecting Short-Run Avoided-Cost Projection

Parameter	Low (%)	Nominal (%)	High (%)
Probability of normal year	0	50	100
Ratio of wet to dry years	0	75	100
South Platte on-margin probabilities: normal year	0	33	100
On-margin probability ratio: Roberts to Moffat, normal year	0	50	100
South Platte on-margin probabilities: wet year	0	75	100
On-margin probability ratio: Roberts to Moffat, wet year	0	50	100
South Platte on-margin probabilities: dry year	0	30	100
On-margin probability ratio: Roberts to Moffat, dry year	0	50	100
Operating-cost escalation	0	0.001	3

all system components. We used a wide range of probabilities to ensure that the true estimated value lies within the range of results calculated. A more detailed analysis would reduce these ranges as justified by agency information and other studies.

Figure 4.2 is a histogram of 1,000 runs drawn quasi-uniformly from the ranges in Table 4.2 and assuming a 6 percent discount rate.[1] The PV of SR avoided costs range from about $7,500 per mgal to $22,000 per mgal. Recall that the nominal factors yielded a result of $10,300 per mgal. The lower-bound estimate is 73 percent of the nominal estimate, and the upper bound is 214 percent of the nominal estimate. In this case, a majority of alternative cases (reflecting different plausible values of the uncertain factors) lead to higher estimates for the SR avoided costs.

Long-Run Avoided Costs

LR cost savings will accrue to Denver Water if efficiency leads to deferral or downsizing of the Moffat Project (18 taf per year or 16.1 mgal per year) or the second, unspecified project (31 taf

Figure 4.2
Histogram of Denver Water System Short-Run Avoided Costs Under Hydrologic, On-Margin Probability, and Cost-Escalation Uncertainty (1,000 runs)

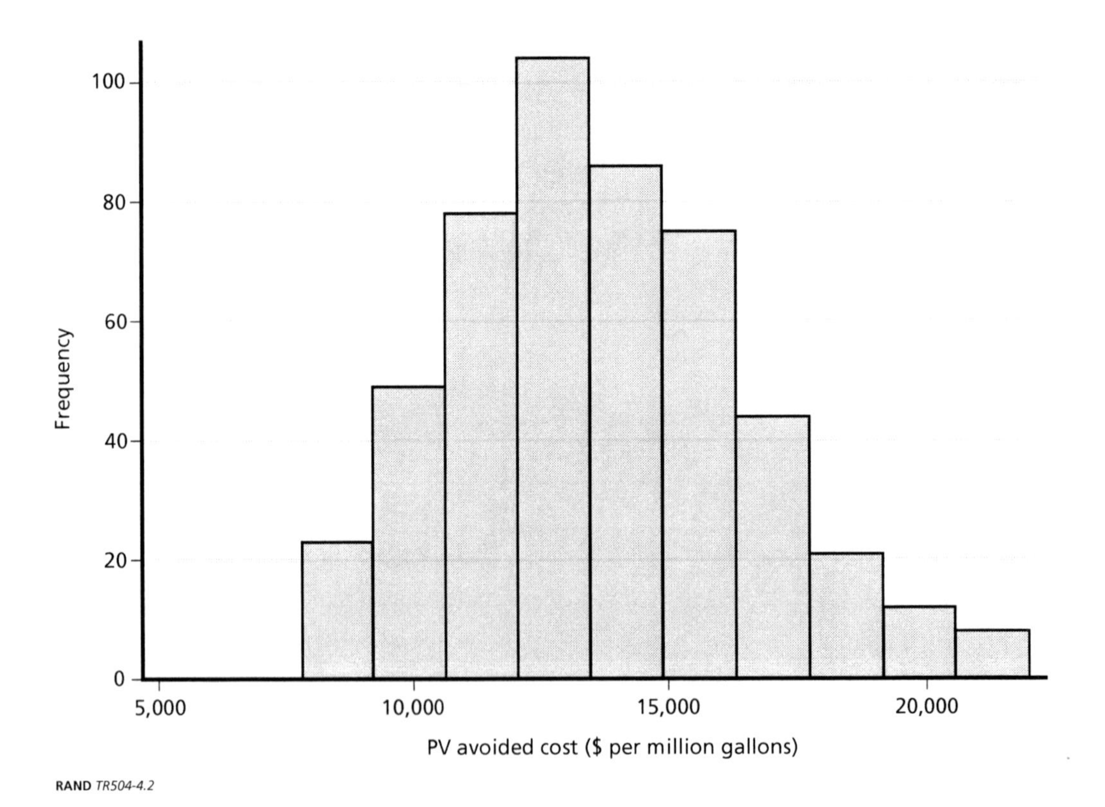

RAND *TR504-4.2*

---

1 We use a Latin hypercube sampling scheme to randomly choose values for the uncertain parameters across their plausible ranges while ensuring that the sample is reasonably uniform.

per year or 27.7 mgal per year). For this analysis, we assume that water-demand reductions would lead these projects to be deferred and not downsized.

Recall that we modified the AC model to consider trends in water demand and supply to estimate the deferral period due to any demand reductions from efficiency. Considering the baseline supply and demand projection, we now evaluate the LR avoided cost. Figure 4.3 shows both the SR and LR avoided costs over time (dollars per mgal) for base-case parameter values. The figure shows two large accruals of LR avoided costs just after 2015 and 2030, pertaining to the cost savings due to the deferral of the Moffat and second projects. Note that the LR avoided costs are considerably larger than the SR avoided costs after 2015.

Just as uncertainty about parameters relevant to the SR avoided costs led to a range of estimates for SR avoided costs, several uncertainties affect the LR avoided costs. Table 4.3 shows four additional key uncertain parameters along with the ranges of values considered in the analysis. The first two control uncertainty in supply and demand as discussed above. The third and fourth parameters pertain to estimates of future project costs. Our estimates of project costs are highly uncertain, since cost estimates for the Moffat Project have not yet been determined and the second project has yet to be identified. For base-case costs, we use estimates from a Bureau of Reclamation study on the cost of developing new water supply in the Front Range of the Rocky Mountains (Bureau of Reclamation and Colorado Springs Utilities, 2004). The study compares projected costs of several proposed projects and a recently

Figure 4.3
Short-Run and Long-Run Denver Water System Avoided Costs (discounted at 6%)

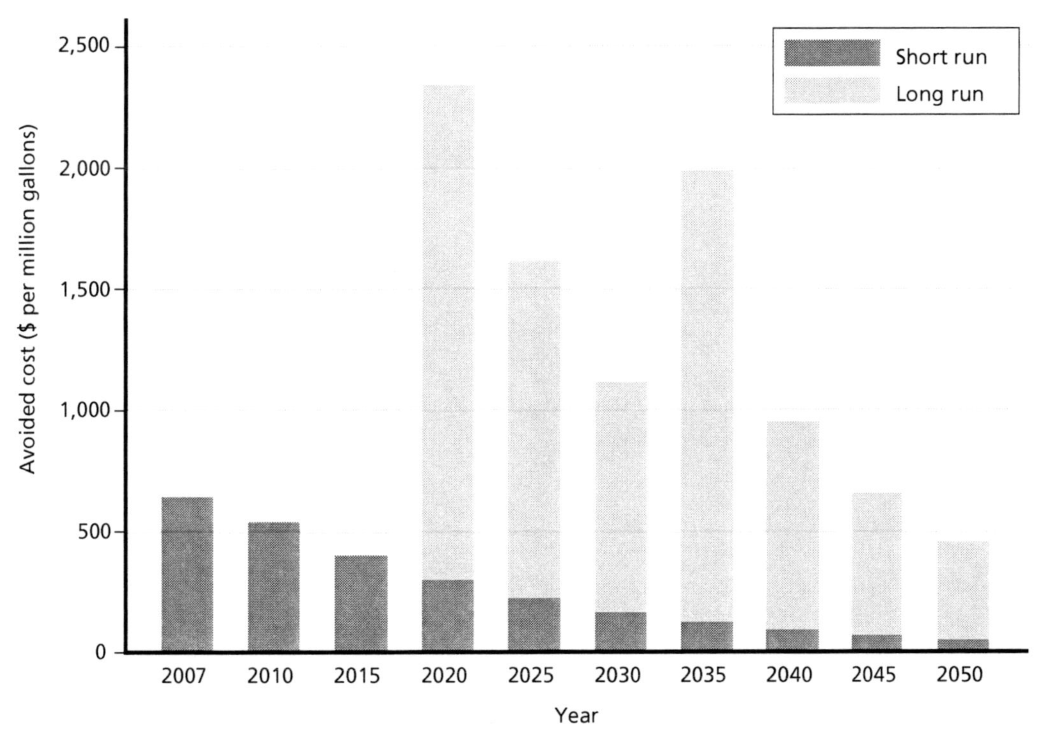

RAND *TR504-4.3*

Table 4.3
Additional Parameters Affecting Long-Run Avoided Cost Projection

Parameter	Low (%)	Nominal (%)	High (%)
Supply decline	–10	0	0
Demand-growth rate change	–5	0	5
Moffat supply-cost uncertainty	0	0[a]	25
Second project supply-cost uncertainty	0	0[a]	25

[a] The base-case cost per capacity of the Moffat and second projects is $64,450 per mgal ($21,000 per AF) (Bureau of Reclamation and Colorado Springs Utilities, 2004).

completed comparable project (Colorado–Big Thompson project). We used the average of these two estimates as our base-case cost estimate.

We again evaluate the models 1,000 times, drawing quasi-uniformly from the uncertain parameters in Table 4.2 and Table 4.3. Note that most of the uncertainties that affect the SR avoided costs also affect the LR avoided costs. LR avoided costs range from about $24,000 per mgal to $51,000 per mgal. Summing the SR and LR avoided costs increases the range of valuations of efficiency benefits to between $36,000 per mgal and $69,000 per mgal, with a median result of about $45,000 per mgal. Figure 4.4 shows the ranges of results for the SR avoided cost only, LR avoided cost, and total avoided cost as a box plot.

A key finding at this point of the analysis is that, despite the significant uncertainty in these projections, the lower-bound estimate for the total avoided costs is significantly higher than the maximum estimate of the SR costs alone (about 70 percent higher). This increase in value attributed to efficiency could be conservatively factored into an agency's planning.

Environmental and Recreational Benefits

We next evaluate the efficiency benefits that accrue to the environment and recreational services in the source regions of the Denver Water supply.

Short-Run Environmental and Recreational Benefits

We consider the following four SR environmental and recreational benefits:

- avoided riparian and wetland habitat impacts
- avoided impacts to river fisheries
- avoided impacts of NO_X emissions
- avoided nonangling recreational impacts.

As described in Chapter Two, a different methodology is used to quantify each benefit. As with the avoided-cost estimates, we run the EB model 1,000 times, sampling from the parameter ranges shown in Table 4.4.

For riparian areas and wetlands, the model multiplies the estimated marginal impact of water extractions on viable habitat to the value of riparian habitat. In the model, the former is influenced by the average river buffer width. We vary this parameter between 25 feet and 75 feet to reflect uncertainty about the marginal impact. We also consider a wide range of

Figure 4.4
Present Value of Short-Run, Long-Run, and Total Avoided Costs

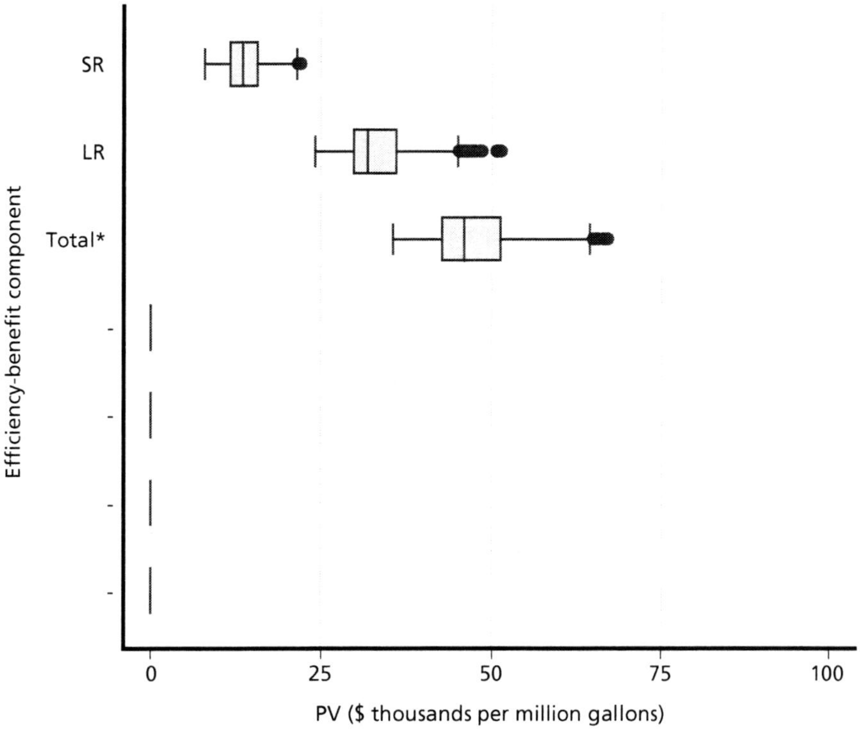

NOTE: Asterisk indicates summation estimates. Each box represents the results of 1,000 model runs and shows the lower quartile (left edge), upper quartile (right edge), and median (inner line) of the results. The dots and stems depict the remaining range of the results.

RAND *TR504-4.4*

Table 4.4
Key Uncertain Environmental-Benefit Model Parameters and Ranges Used for Analysis

Parameter	Low	Nominal	High
Average river buffer width (ft)	25	50	75
Value of riparian habitat ($/acre)	215	500	1,000
South Platte percentage of statewide fishing activity (%)	0	2.5	5
Electricity emission rate (lbs of NO_x/kwh)	0.001	0.002	0.005
NO_x emission valuation ($/ton)	1,487.5	6,360	6,360
Marginal rafting impact (user-days/mgal)	0.0561	0.0835	0.1109
Value of rafting ($/user-day)	100	267	500

riparian habitat valuations bounded by the annualized average purchase price for California riparian habitat ($215 per acre) and a higher value ($1,000 per acre) to reflect the inclusion of wetlands, which, in California, are priced about 10 times that of riparian habitat ($2,500 per acre).

For fishing, the model estimates the economic impact on fishing from water use by apportioning the marginal change in statewide fishing observed between 2001 and 2003 to the South Platte, Blue, and Fraser rivers and then using flow differences between these years to estimate the marginal impact. To reflect uncertainty about the marginal impact, we varied the percentage of total Colorado recreational fishing attributable to the South Platte River (from 0 percent to 5 percent), as shown in Table 4.4.[2] The economic impacts of the other two rivers are then adjusted to maintain impacts that are proportional to the relative river lengths.

The model estimates the economic impact of NO_X emissions by multiplying estimates of the marginal NO_X emissions by the value of those emissions. To reflect uncertainty about the marginal emissions, we vary the amount of NO_X emitted per kwh of electricity (from 0.001 lbs per kwh to 0.005 lbs per kwh). We used a wide range of NO_X emission valuations ($1,487 per ton to $6,350 per ton), as shown in Table 4.4.

Finally, for nonangling recreation, the model multiplies the estimated marginal impact of water extractions from the Blue River on the number of rafting days by the estimated economic impact of a person-day of rafting.[3] The ranges for the marginal rafting impact are the 95 percent confidence interval for the regression coefficient used as the nominal value. The range for the rafting valuation was chosen arbitrarily to reflect significant uncertainty about the estimated economic impact value. Table 4.4 shows the range of values used.

Table 4.5 shows the ranges of benefit estimates for each of the four environmental services evaluated. For nonangling recreation, riparian habitat, and air emissions, the estimates of annual marginal benefit of water-use reductions are very low. This result could be for several reasons. First, the benefit to efficiency to these services could actually be low in the Denver Water area. Second, the methodologies used could be based on unrealistically low impact or valuation estimates. Finally, the particular benefit estimates could be reasonable but do not include other related benefits that could be larger, such as those to lake recreation. Resolving these issues is beyond the scope of this study.

Table 4.5
Range of Environmental and Recreational Benefits for Each of the Four Environmental Services (for 1,000 model runs)

	Environmental and Recreational Benefits ($/mgal)	
Service	**Minimum**	**Maximum**
Riparian and wetlands	1	30
Fish	0	43,700
NO_X	1	49
Nonangling recreation	1	357

[2] The percentage range of fishing activity attributable to the South Platte River used in the analysis (0 to 5 percent) may be conservative. The South Platte River is the only major river flowing through Denver, Jefferson, and Park counties, and economic activity attributable to the fishing industry in these counties accounts for more than one-third of the total statewide fishing economic activity (BBC Research and Consulting and Colorado Division of Wildlife, 2004, 2002 data). Because some of this economic activity likely supports fishing in rivers in other counties, we chose a very conservative estimate of commercial activity of 0 to 5 percent.

[3] Note that no significant relationship was found between the South Platte River rafting days and river flow and that no recreational-use data were available for the Fraser River.

The fishery benefits, on the other hand, are significant. The river-fishing benefit ranges from zero benefit to about $44,700 per mgal. These results derive from the large amount of economic activity attributable to river fishing in the rivers impacted by Denver-region water use. It is difficult to know, however, how much of this statewide economic activity would be impacted by changing flows in one or only a few rivers in Colorado. This method assumes that increased flows due to efficiency will lead to some marginal increase in economic activity related to fishing, but it is unknown whether this would come at the expense of other rivers and streams in Colorado. Alternative methodologies would likely provide significantly different results. For example, nonmarket economic-valuation studies that assessed the value of recreational benefits from fishing and rafting in Colorado rivers showed much larger levels of benefits (Platt, 2001). These studies have many known weaknesses and potential biases, such as one's tendency to overstate one's valuation of a good in hypothetical situations; however, we use these studies not as a basis of our estimates but only to illustrate that our estimates remain in a plausible range.

Long-Run Environmental and Recreational Benefits

The CUWCC EB model calculates only environmental benefits associated with reduced extractions of water due to efficiency and not any benefits associated with the deferral of water-supply projects that have environmental effects. We thus augment the SR environmental benefits estimated by the EB model by adding an additional category of environmental costs to the AC model. The deferral of these monetized environmental costs constitutes a legitimate benefit of conservation. Note that this benefit could be quite large if projects were completely eliminated due to conservation—although, in this context, this is very unlikely.

For this study, we could not identify any studies that provided quantified environmental impacts due to the Moffat Project, and impact estimates for the undefined, second project were also unavailable.[4] Major concerns revealed during the initial scoping of the Moffat Project, however, included the proposed diversions on the health and functioning of the Fraser River and on the "water-related, recreation-based" economy of the Fraser River Basin, particularly Grand County (USACE, 2003). Therefore, we consider a wide range of possible ongoing impacts of the project from $0.5 million per year to $5 million per year. For the second project, we assumed a range of $1 million per year to $10 million per year.

Figure 4.5 shows all the avoided costs and included environmental and recreational benefits for the Denver Water system. SR environmental and recreational benefits range from about $103 per mgal to $43,700 per mgal. LR environmental benefits range from $900 per mgal to about $8,600 per mgal. Together, the environmental and recreational benefits include a significant amount of value to efficiency programs. Note that the median estimate of total environmental and recreational benefit ($15,300 per mgal) is about the same as the median SR avoided-cost estimate ($13,400 per mgal), which is similar to the value typically ascribed to efficiency programs—adding the environmental and recreational benefits represents significant increases in efficiency value. Furthermore, these coarse estimates of environmental benefits likely do not capture all environmental benefits. Agencies thus should strongly consider estimating these benefits when valuing efficiency programs.

---

[4] The draft environmental impact statement is due out in October 2008, and the final version in May 2008.

Figure 4.5
Present Value of Short-Run, Long-Run, and Total Avoided Costs and Short-Run, Long-Run, and Total Environmental and Recreational Benefits

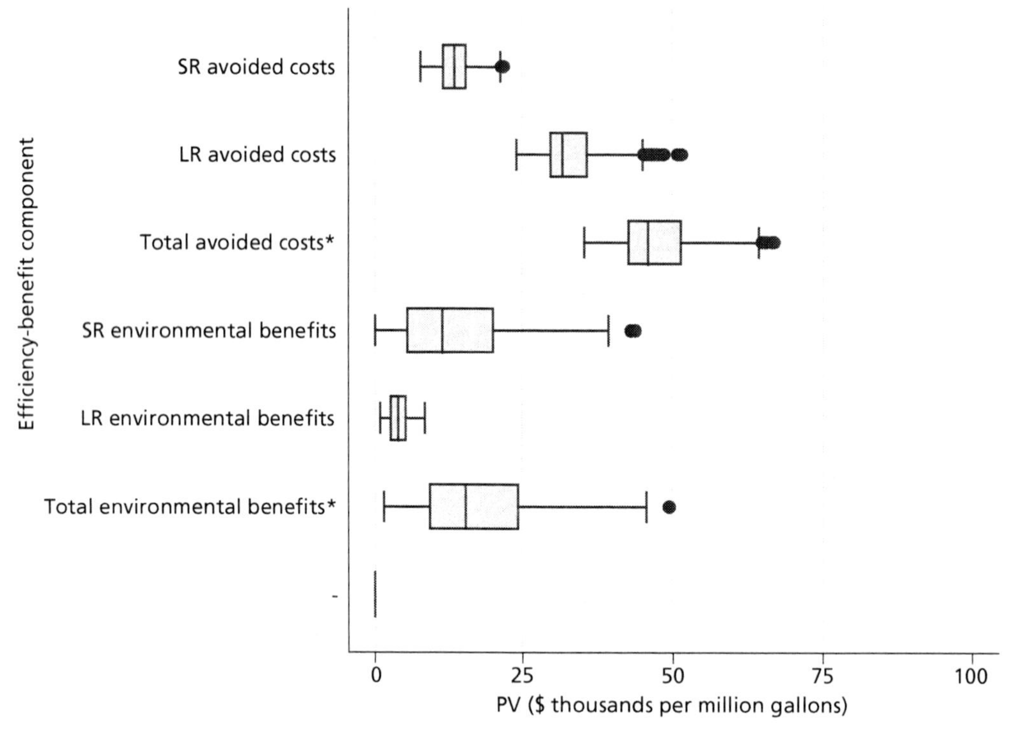

NOTE: Asterisks indicate summation estimates. Each box represents the results of 1,000 model runs and shows the lower quartile (left edge), upper quartile (right edge), and median (inner line) of the results. The dots and stems depict the remaining range of the results.
RAND *TR504-4.5*

Total Benefits

The total efficiency benefit estimated is the sum of the avoided costs and environmental and recreational benefits. Figure 4.6 adds the final result to those of the individual components in Figure 4.5. After including all avoided costs and environmental and recreational benefits (right-most box), the estimates suggest a range of marginal benefits between about $41,000 per mgal and just about $100,000 per mgal. Although the range of this estimate is quite large, the lower bound of the range ($41,000 per mgal) is about double that of the upper range of the SR avoided costs ($22,000 per mgal). This suggests that, if an agency were to consider only those costs, it could be undervaluing efficiency by 50 percent or more. Even with coarse data and rough methodologies, it can be useful to quantify the LR avoided costs and the environmental and recreational benefits.

These estimates also assume that efficiency savings are uniform throughout the year. Recall that the magnitude of LR avoided costs depends only on water savings during peak months. An efficiency program that saves more water during the peak season (when it is hot or dry) will have greater benefits per unit water saved, since all the efficiency savings will be

Figure 4.6
Present Value of Short-Run, Long-Run, and Total Avoided Costs; Short-Run, Long-Run, and Total Environmental and Recreational Benefits; and the Sum of the Total Avoided Costs and the Total Environmental and Recreational Benefits

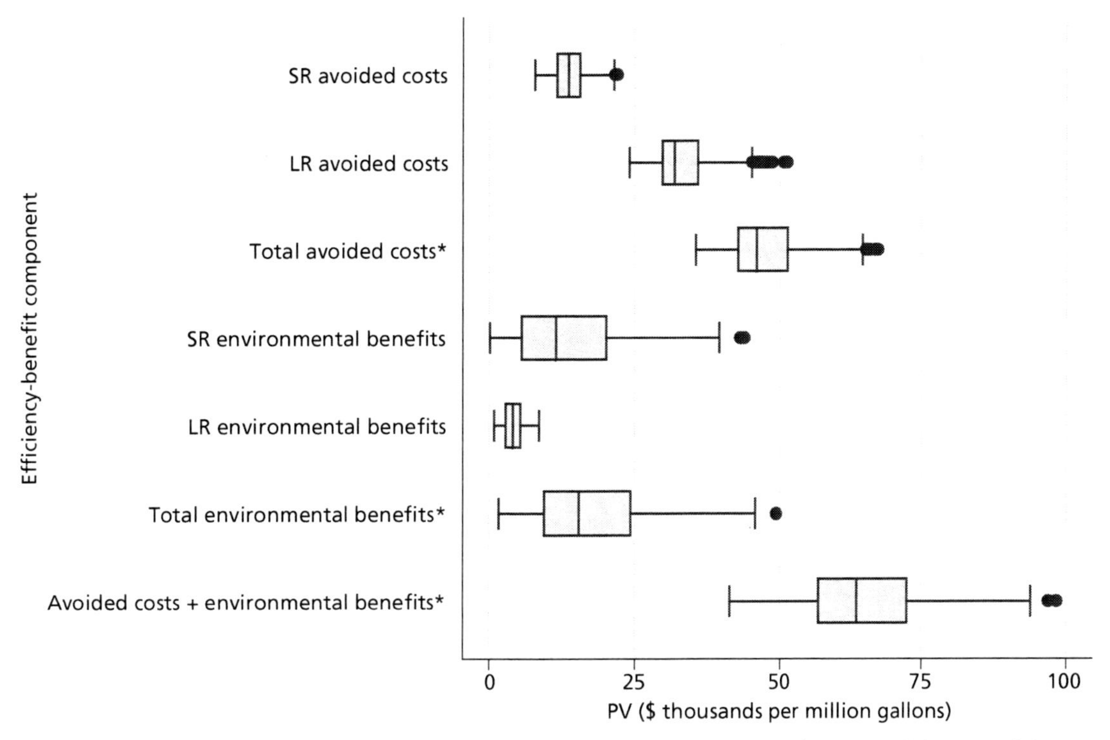

NOTE: Asterisks indicate summation estimates. Each box represents the results of 1,000 model runs and shows the lower quartile (left edge), upper quartile (right edge), and median (inner line) of the results. The dots and stems depict the remaining range of the results.
RAND *TR504-4.6*

concentrated during the months when the LR benefits accrue. Figure 4.7 compares the calculated efficiency benefits for a program that saves water uniformly with one that saves only during peak months. As can be seen in Figure 4.7, programs that save water primarily during peak periods, such as an outdoor efficiency program, have much larger benefits. In this case, benefits range between $81,000 per mgal per year and $170,000 per mgal per year.

Evaluating Efficiency Programs

We now compare these marginal benefit estimates to cost estimates for efficiency measures from Denver Water's proposed 10-year conservation program (Denver Water, 2006b). Table 4.6 lists 21 different measures included in the conservation program. The first column indicates the type of measure. The second column provides the measure names and indicates those that would lead to savings primarily during peak periods. The third column indicates the projected 10-year undiscounted project costs, which includes costs borne by the utility and the end users. The fourth column indicates the average annual water saving estimates, calculated by dividing the reported 10-year savings amount by 10 years.

Figure 4.7
Comparison of the Present Value of Benefits, Based on Total Avoided Costs and Total Avoided Costs Plus Environmental Benefits, for Efficiency Programs That Save Water Uniformly Throughout the Year and Only During Peak Periods

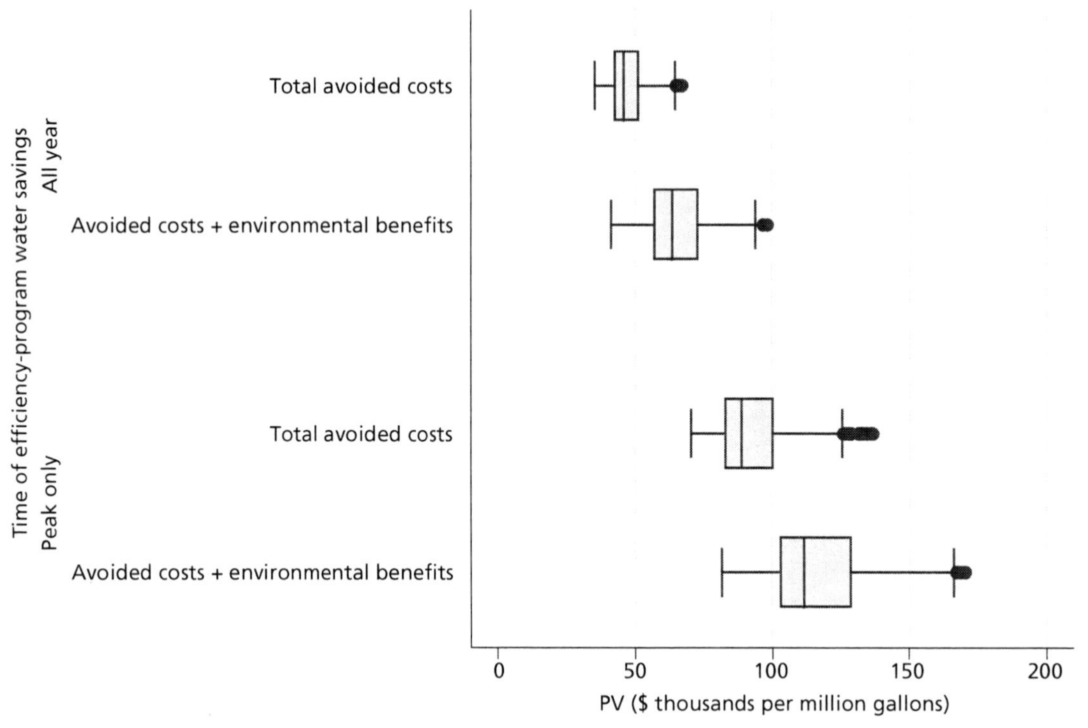

RAND *TR504-4.7*

Table 4.6 shows that the total cost of Denver Water's programs vary widely (from $80,000 for a low-flow urinal–replacement program to more than $109 million for an irrigation-efficiency incentive program. Savings also vary widely from 20 mgal per yr for a car-wash certification program to 2,280 mgal per yr for a time-of-sale retrofit program for toilets, showerheads, and faucets.

To help visualize how the measures compare in terms of costs and savings, Figure 4.8 plots the projected 10-year total costs (undiscounted) against the water savings for each of the programs listed in Table 4.6. The right y axis indicates the PV of discounted total spending if the program were extended from 2007 to 2050 (see Table D.1 for data). Those programs to the lower right of the plot save more and cost less than those programs to the upper left. For example, the conservation education program is projected to save considerably more water per level of expenditure than will the irrigation checkup program.

The average total program costs and average saving information can be used to estimate the average cost of water savings from an individual measure. To evaluate cost effectiveness of efficiency programs using the marginal benefit estimates from Table 4.6 and Figure 4.8, however, one needs estimates of the marginal costs of the efficiency programs. Although this is not possible to do given the data available, we can speculate about the relationship between average costs and marginal costs. If marginal costs are constant, then marginal and average costs are the same. For example, for a low-flow toilet rebate program, the marginal cost of implementing the program is likely to remain constant over a wide range of rebates, and having constant

Table 4.6
Efficiency Program Measures, Projected Total Cost, and Yearly Water Savings in Denver Water's Proposed 10-Year Conservation Plan

Type	Efficiency Measure	Projected 10-Year Total Cost ($ thousands)	Water Savings (mgal/yr)[a]
E	Car-wash certifications	110	20
E	Xeriscape planning and design	10,820	20
E	Multifamily residential audit program	490	20
R/I	Low-flow toilet	250	40
R/I	Public housing retrofits	3,540	50
R/I	Wireless rainfall sensor rebate[b]	760	50
E	Conservation kiosks	1,200	90
R	Low-flow urinal requirement	80	100
R/I	High-efficiency toilet rebate	4,290	130
R	Irrigation-meter requirement[b]	2,170	210
R/I	Evapotranspiration (ET) controller rebate[b]	410	230
R/I	Clothing-washer rebate	24,600	440
E	Irrigation classes and seminars[b]	3,740	470
E	Cooling-tower monitoring[b]	870	520
R/I	Natural-area conversion for large landscapes[b]	22,580	560
E	Irrigation checkups for large irrigators[b]	35,810	620
R/I	Irrigation-efficiency incentives[b]	109,090	680
E	Conservation education program	4,570	960
R	Water-efficiency rating for new customers	27,580	1,240
R/I	Commercial and industrial incentives	76,320	1,520
R	Time-of-sale retrofit of toilets, showerheads, and faucets	49,110	2,280

SOURCE: Denver Water (2006b).

[a] Computed by the authors.

[b] Measure that is likely to concentrate savings during peak months.

NOTE: E = educational. R/I = rebate or incentive. R = regulatory.

marginal costs thus may be a satisfactory assumption. Many of the regulatory and educational programs, in contrast, are likely to have increasing marginal costs as the programs scale up. For instance, for an auditing program, early participants are likely to be those who also will realize large water savings in response to the audit. As the program continues, the water savings per audit may drop, resulting in rising marginal costs. In these cases, the average costs calculated in Table 4.6 understate the marginal costs. Lastly, agencies could become more efficient at implementing a program as it grows through "learning by doing." In such cases, marginal costs would decrease as savings are realized.

Figure 4.8
Water Savings Versus 10-year Total Program Cost for Each of Denver Water's Proposed Efficiency Programs

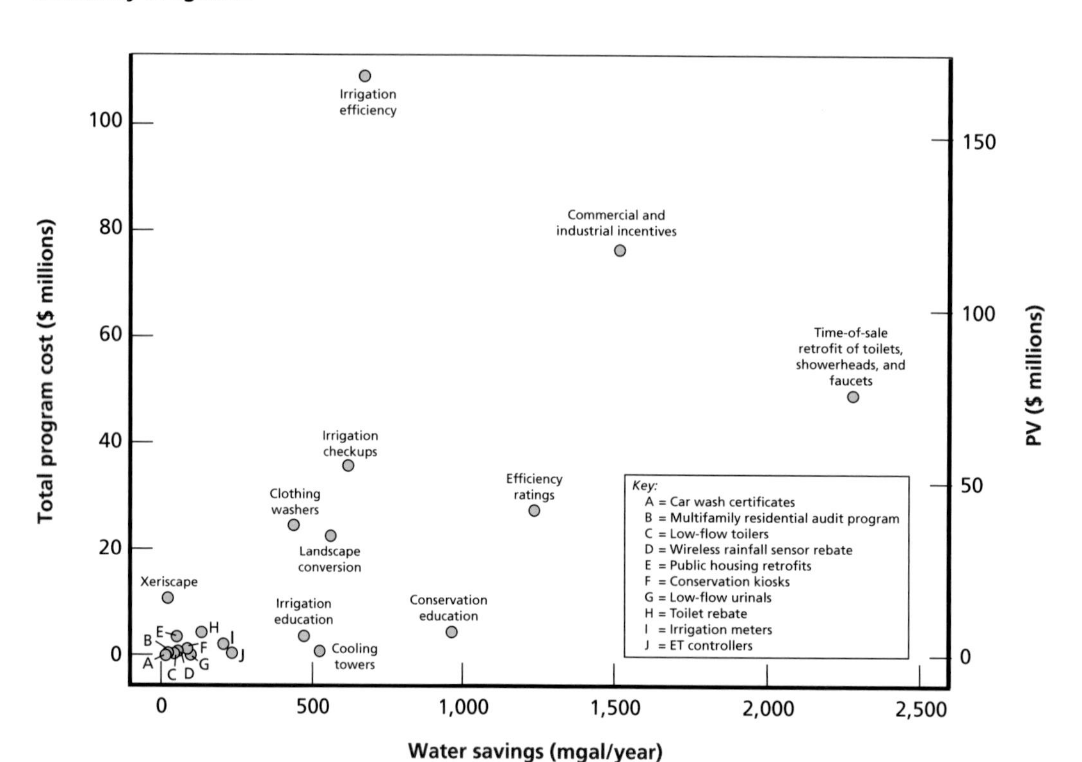

NOTE: The right y axis indicates the PV of total spending if the program were extend from 2007 to 2050.
RAND *TR504-4.8*

These conditions are not mutually exclusive, and a program may exhibit all three at some point in its lifetime. In a retrofit program, for example, the first few retrofits may be costly, but the marginal costs decline as the program becomes more efficient. After this initial phase, marginal costs may be constant, since additional retrofits take a similar level of effort to achieve the same water savings. As the program expands, though, the program will have utilized the easy opportunities, and marginal costs rise as the retrofits become more difficult.

It is beyond the scope of this study to reconcile the differences between the average costs that can be computed using the data in Table 4.6 and the marginal cost for each program. An agency, however, should be able to make these assessments for its programs without too much difficulty. For the remainder of this discussion, we assume that marginal costs equal average costs.

Comparing Efficiency Benefits to Efficiency-Program Costs

Table 4.7 summarizes the minimum and median values of efficiency benefit estimates for Denver Water, assuming uniform water savings and water savings only during peak months. Recall that considering only the SR avoided costs yields a minimum and median value of efficiency of $7,800 per mgal and $13,400 per mgal, respectively. The benefits are much larger when LR avoided costs and environmental and recreational benefits are included (a median estimate of $63,400 per mgal for uniform savings) and for programs that save water only during peak periods (median estimate of $111,600 per mgal).

Table 4.7
Minimum and Median Values for Estimated Total Efficiency Benefit to Denver Water, Assuming Uniform Savings and Peak-Only Savings

Efficiency-Benefit Components	Minimum Value ($/mgal)	Median Value ($/mgal)
Uniform savings		
SR avoided costs	7,800	13,400
SR and LR avoided costs	35,400	45,800
Avoided costs + environmental benefits	41,200	63,400
Peak-only savings		
SR avoided costs	7,800	13,400
SR and LR avoided costs	70,200	88,800
Avoided costs + environmental benefits	81,400	111,600

Figure 4.9
Efficiency Programs That Save Water Uniformly Throughout the Year, Ranked by Average Cost

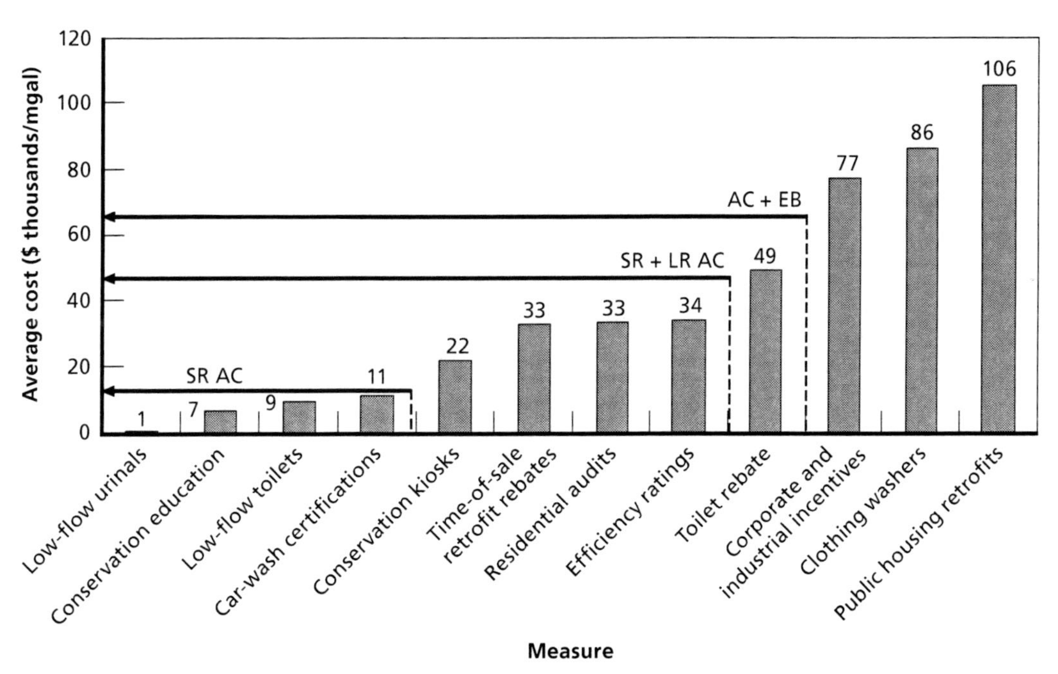

NOTE: The vertical bars show the average cost of each efficiency program. The horizontal arrows are positioned to demarcate the range of programs that are cost-effective when considering the various benefits. The height of the arrow corresponds to the median efficiency-valuation estimate. These arrows indicate which programs have average costs below the calculated median marginal-benefit estimates, assuming that savings accrue uniformly throughout the year.
RAND *TR504-4.9*

We next combine the average Denver Water efficiency-program cost information (from Table 4.6) with the minimum and median estimates of marginal efficiency benefits to Denver Water (from Table 4.7). Figure 4.9 shows this comparison for programs that reduce demand uniformly throughout the year. Making the assumption that the marginal costs of the Denver

Water efficiency programs are equivalent to average costs, any program with average costs lower than the marginal benefit of efficiency is cost-effective from an economic perspective. For example, if Denver Water were to consider only SR environmental benefits, then only the four lowest-cost programs (low-flow urinals through car-wash certifications) would be cost-effective (less than the median value of benefits). If the minimum value of benefits were considered, then only the first two programs would be cost-effective. By including both the avoided costs and environmental and recreational benefits, all uniform-saving programs except for corporate and industrial incentives, clothing-washer rebates, and public housing retrofits are cost-effective (using the median benefit value). Using the minimum total benefit value, only one less program would not be cost-effective: the toilet rebate program.

Figure 4.10 shows similar information to that in Figure 4.9 but only for programs that reduce demand primarily during the summer months. In this case, only three of the nine programs are cost-effective if only SR avoided costs are considered. Including LR avoided costs, however, leads to four more cost-effective programs, including irrigation checkups, that are estimated to cost $89,000 per mgal (SR + LR avoided costs). Including environmental and recreational benefits (avoided costs + environmental benefits) does not justify any more of the currently considered programs, because of the large increase in cost for the next-most-expensive efficiency program: irrigation efficiency. This does not imply that the environmental and recreational benefits are not important to consider; in fact, these additional benefits suggest that a program that costs 24 percent more than the irrigation checkup program would still be cost-effective, as would be additional investment in the lower-cost programs. Figure 4.10 would change only slightly if one considered the lowest estimate for each benefit; only one more program would be estimated as not cost-effective: irrigation checkups.

Figures 4.9 and 4.10 also suggest where agencies may choose to expand their efficiency programs. For example, the average cost per water savings for the low-flow toilet rebate program is very low ($9,500 per mgal) compared to the estimated marginal benefit of the efficiency when including the LR avoided costs and environmental and recreational benefits ($62,900 per mgal), suggesting that an expansion of this program, even at increasing marginal costs, would be cost-effective.

Figure 4.10
Efficiency Programs That Save Water Primarily During Summer Months, Ranked by Average Cost

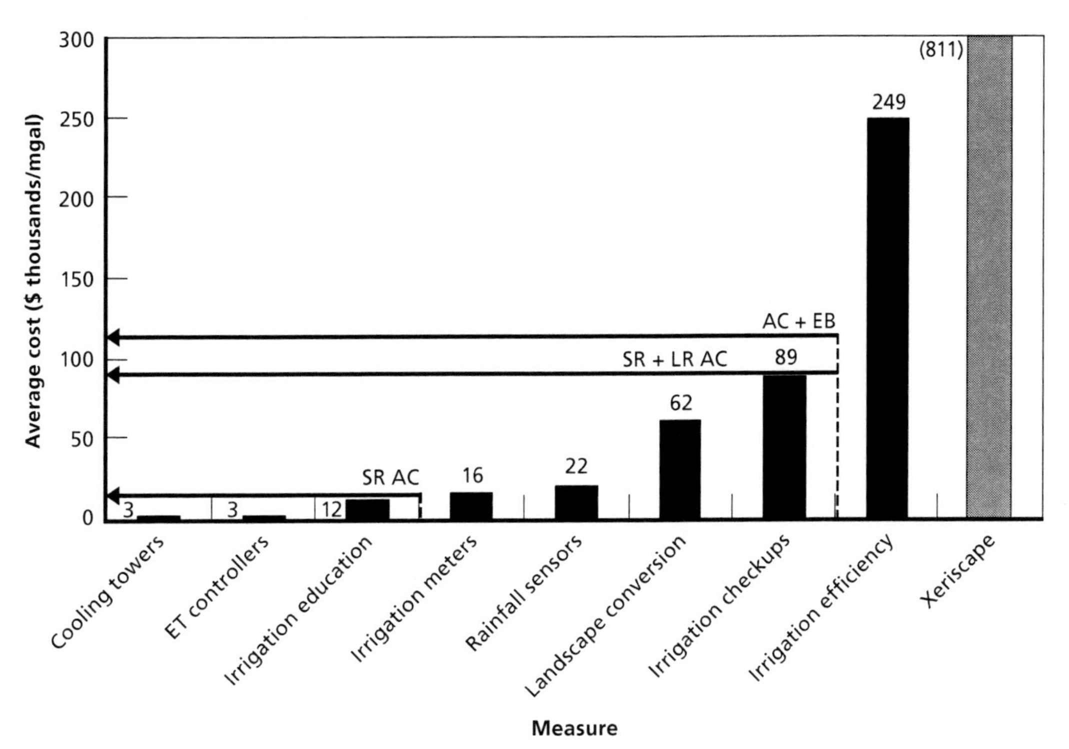

NOTE: Arrows indicate which programs have average costs below the calculated median marginal benefit estimates, assuming that savings accrue during the peak months.

RAND *TR504-4.10*

CHAPTER FIVE

Summary and Conclusions

It is common practice for water planners to use simple cost-effectiveness heuristics when choosing how much investment to make in reducing water demand through water-use efficiency. Often, these simple methods consider only the cost of acquiring additional supply from existing systems. Because there are other costs borne by the utility and by society that are ignored, this method leads to undervaluation of water-use efficiency.

In this report, we presented an economic framework based largely on recent work done for the CUWCC for evaluating the marginal benefit of water-use efficiency savings to a water utility. This framework considers additional benefits over the typically considered cost savings from reducing the water supply delivered to the end user (SR avoided costs). The framework also includes benefits from deferring or downsizing future supply-related infrastructure projects (LR avoided costs), benefits from reducing water extractions from natural systems that support environmental and recreational services (SR environmental benefits), and benefits from deferring water-supply projects that would have sizable environmental and recreational impacts. The sum of these benefits comprises a more comprehensive estimate of the marginal benefit of water-use efficiency. This estimate can then be used to identify which water-use efficiency programs are cost-effective or to design programs that maximize water savings within a cost-effectiveness and budget constraint.

We illustrated this framework through a case study focused on the Denver Water service area by adapting two models recently developed for the CUWCC to the Denver Water service area. Specifically, we developed a simplified representation of the Denver Water system for use with the CUWCC AC model. Also, we re-estimated environmental and recreational valuations and water-use impacts to use the CUWCC EB model for the Denver service area. The analysis uses exploratory modeling techniques to develop a wide range of plausible results. Denver Water was not a formal partner in this study, and we used publicly available data. As such, the assessments of these benefits are suggestive and not conclusive.

Even though the results are of low precision, they provide several key findings. First, if one considers *only* the SR avoided costs, then many water-efficiency projects included in Denver Water's 10-year conservation plan would not be estimated to be cost-effective. For example, the median estimate of efficiency value when considering only SR avoided costs is $13,400 per mgal of water savings (for programs that save water uniformly throughout the year). Only four of the 12 Denver Water efficiency programs cost less than this on average. When LR avoided costs and efficiency and recreational benefits are estimated and added to the marginal-benefit calculation, an additional five programs are cost-effective (based on the median result). All but three Denver Water programs (with uniform savings) were estimated to be cost-effective using this efficiency valuation. If Denver Water had used strictly a cost-effectiveness criterion based solely on SR avoided costs, it would likely have chosen to implement only a small number of efficiency programs.

This framework also highlights the importance of considering the annual saving profile of water-efficiency savings. Water savings from programs that concentrate savings during summer months should be valued higher than savings from programs that lead to more uniform water savings throughout the year. This is because many agencies make capital investments to increase water supplies in response to rising peak demand, and efficiency programs that save water exclusively during peak times will lead to greater deferral periods of supply projects than will efficiency programs whose savings are spread out throughout the year. We identified such efficiency programs in Denver Water's 10-year conservation plan and applied efficiency valuations that reflect this increased efficiency value. For our case study, the median value of efficiency benefits for a peak-saving efficiency program was 76 percent higher ($111,600 per mgal) than for a year-round efficiency program.

Our final analytic finding focused on uncertainty. We evaluated our models 1,000 times to sample over a wide range of plausible values for uncertain inputs. The range of inputs chosen included very conservative values, so the minimum values evaluated using this method are also conservative. Notably, even this broad range of outcomes does not change the analysis considerably. The lower estimate still suggests that most of the Denver Water programs are cost-effective and that significantly more investment could be made to expand those programs that are cost-effective.

Our case study demonstrated that this framework and these models can be successfully applied to regions outside of California. An investment is required to calibrate the model's parameters to local circumstances, although water agencies with access to better data and information about their water systems will be able to do so more easily and more accurately. Obtaining better estimates of the environmental impacts of existing water use will also dramatically improve the assessments of the environmental and recreational benefits. Together, this would lead agencies themselves to produce efficiency-valuation estimates with smaller ranges and likely lead to higher confidence in the cost-effectiveness of more water-use efficiency programs.

APPENDIX A

Avoided-Cost Model

The CUWCC AC model assumes an incremental increase of 1 million gallons of water conservation each year and then calculates the utility's avoided costs in the SR and LR from this increase in water conservation. The SR avoided costs accrue by saving expenditures on operating the utility's water supply, treatment, and delivery systems. LR savings result from deferring or downsizing future supply projects. The model sums the SR and LR costs for each year in the planning horizon to estimate total avoided costs each year. This appendix provides a more detailed description of the AC model, discusses how Denver Water's system is represented in the model, and explains modifications to the model made for this application to Denver Water.

Description of the Avoided-Cost Model

The model uses data on demand, operating costs, and future supply costs to calculate the SR and LR avoided costs. The model assumes an incremental increase in water conservation (1 million gallons) and calculates the expected savings from reducing system operation for the SR avoided costs. In the LR, the model estimates the change in annualized costs after deferring or downsizing future supply projects through conservation. This method attempts to account for varying operating costs among different parts of the utility's system and potential operational constraints on system components.

The model has the following main inputs:

- *water-demand schedule*: projection of the rate of water demand in each year of the planning horizon for the peak and off-peak periods
- *existing system components*: description of the main components of the water system, including its system function (supply, storage, treatment, or conveyance), online year for future projects, and water-loss rate
- *operating costs for system components*: estimated operating costs for each system component
- *on-margin probabilities for system components*: estimated probability that each component is the marginal water supply
- *proposed future supply projects*: description of proposed future water-supply projects, whether they can be deferred or downsized, and projected online dates
- *cost estimates for future supply projects*: estimated O&M costs and capital costs for future projects.

In addition to these main inputs, the user is required to specify the planning time horizon, discount rate, real escalation rate for operating costs, peak and off-peak periods, and average water-loss rate. The model uses all these inputs in the SR and LR avoided-cost estimates.

Short-Run Avoided-Cost Estimation

The model estimates SR avoided costs as the expected savings from reducing operating costs. The logic behind the calculation is that an incremental increase in water conservation will reduce water use at the marginal water supply (the system component used to supply the last unit of water). The model combines the user-supplied data on system components, their operating costs, and on-margin probabilities to calculate the expected savings on operating costs. Formally,

$$\text{SR avoided cost} = \sum_{i} \sum_{j} \text{on-margin prob}_{ij} \times \text{operating cost}_{ij},$$

where i is a system function = (supply, storage, treatment, conveyance) and j is a system component. The model allows users to specify different sets of on-margin probabilities for dry, wet, and normal years, and then makes the expected-value calculation with user-assigned probabilities for the different climatic conditions.

Long-Run Avoided-Cost Estimation

The model estimates LR avoided costs by calculating the difference in annualized project costs with and without the increase in water conservation. The logic underlying this portion of the model is that an increase in water conservation allows the utility to defer or downsize investment in a future water-supply project. The LR avoided cost of the water conservation is the change in the PV of the project after it has been deferred or downsized.

The user specifies whether a project can be deferred or downsized. If a project can be deferred, the model calculates a deferral period based on peak-period demand. The deferral period is the amount of conservation divided by peak-period demand, which, in this case, simplifies to 1/(peak demand) because the model assumes 1 unit of conservation. This deferral period is the amount of time used to calculate the change in annualized project costs. For downsized projects, the user specifies the fraction of the project that can be downsized from conservation. The model then calculates the change in annualized costs from the differences in the capital costs between the initially proposed project and the smaller project.

Representation of the Denver Water System

System Demand

The AC model requires annual estimates of average daily demand during peak and off-peak periods. We adapted Denver Water's demand projections to fit the model's format. Denver Water projects annual total water demand, and, from these estimates, we calculated peak and off-peak demand using the following steps:

1. Begin with Denver Water's estimate of total annual demand.
2. Calculate average daily demand.

3. Multiply average daily demand by peak and off-peak multipliers.

We estimated the peak and off-peak multipliers using a least-squares fit to Denver Water's projections. That is, we find the peak and off-peak multipliers that minimize the sum of the squared residuals, where the residual is the difference between the annual demand projection from our estimates and Denver Water's projections. Using this method, we calculated a peak-period factor of 1.45 and off-peak factor of 0.68.

System Components

The AC model divides system components into four categories: raw water supply, storage, conveyance, and treatment. For simplicity, we followed Denver Water's broad division of its system into three constituent systems—the Moffat Collection System, Roberts Tunnel Collection System, and South Platte Collection System (see Figure 2.1 in Chapter Two).

We treat each collection system as a separate supply source, storage, and conveyance path. This, of course, is a significant simplification of the system, because each collection system contains multiple raw-water sources and storage facilities. Increased information on the system's operation procedures would permit a more detailed representation in the model. In addition to the supply sources, Denver Water has three water-treatment plants: the Foothills Plant, Marston Plant, and Moffat Plant. Water from the Roberts Tunnel and South Platte collection systems supply the Foothills and Marston plants. The Moffat plant is supplied by water from the Moffat Collection System and is isolated from the two other water-collection systems. This isolation is the primary justification for the proposed water-supply additions in the Moffat Collection System. During periods of severe drought, the Moffat Treatment Plant does not receive sufficient water, and Denver Water may need to shut the system down, as it did in 2002. System planners are concerned that, if an unplanned outage to one of the other plants occurred while the Moffat Plant were inoperable during a drought, the entire system would be seriously jeopardized.

On-Margin Probabilities

The AC model uses estimates of the on-margin probability of each system component to calculate cost savings resulting from a marginal reduction in water supply due to efficiency. For example, if Denver Water reduced water consumption by 1 million gallons, the on-margin probability of a particular component reflects the probability that water use would decrease for that component. For an in-depth analysis, the model documentation suggests developing these probabilities using a detailed simulation model that captures the range of hydrologic conditions and operational constraints imposed on a system.

In our initial representation of the model, we considered on-margin probabilities for three different conditions: wet years, normal years, and dry years. Currently, we have linked the probabilities between the Moffat Collection System components, the Roberts Tunnel supply and Marston Treatment Plant, and South Platte supply and Foothills Treatment Plant. The Moffat components are linked by necessity, because this system operates separately from the others. We have maintained links between the Roberts Tunnel and Marston plants because their probabilities are likely to be correlated. During dry years, the system-operation information in the IRP states that Dillon Reservoir is one of the marginal water supplies. During these periods, demand for treated water increases, which is also likely to raise the probability that the Marston Treatment Plant is a marginal supply, because it is costlier to operate than the Foothills plant. In reality, the Marston and Foothills plants use water from the South Platte and Roberts Tunnel systems. We also test these assumptions on on-margin probabilities.

For this basic analysis, we did not have access to detailed information on Denver Water's system operation, and we made assumptions about the on-margin probabilities under wet, average, and dry conditions. We then used extensive sensitivity analysis to test how important these assumptions are to the model results.

For normal years, we set the on-margin probabilities even for all the systems. While we assume variation in the operating costs that would imply differing on-margin probabilities even in normal years, we do not have enough information on operational constraints and have assumed equal probabilities for the system components under normal conditions. For wet years, we set the on-margin probabilities for the Roberts Tunnel and Moffat systems to be lower because, as water demand decreases, the systems are costlier to operate and available supplies in the South Platte system increase. The on-margin probability for the South Platte system in wet years is higher because it has lower operating costs. In dry years, system demand increases, which requires increased use of the peak-demand treatment plants. For this reason, we have assumed higher probabilities for the Moffat and Roberts Tunnel system while reducing the probability for the South Platte system. Table A.1 displays the probabilities used in the model.

Operating Costs

The third element of the avoided-cost analysis is the operating costs of existing system components. This information is used to calculate the SR savings from conserving water. In our analysis, we used data from the 2005 annual report and 2006 budget (Denver Water, 2005, 2006a) to estimate these costs. These show the 2005 total operating expenses and amounts of water supplied, treated, and delivered. From this information, we calculated the average costs for each supply function. We estimated how these costs varied for different system components. The IRP provides general guidance on variation in costs. For instance, it states that the Foothills Plant is considered the base-load treatment plant, whereas the Marston and Moffat plants are used for peaking. This implies that the Foothills plant has lower operating costs

Table A.1
On-Margin Probability Initial Assumptions

Climate	Water Supply	Treatment Plant	On-Margin Probability
Wet year	South Platte	Foothills	0.5
	Roberts Tunnel	Marston	0.25
	Moffat	Moffat	0.25
Normal year	South Platte	Foothills	0.33
	Roberts Tunnel	Marston	0.33
	Moffat	Moffat	0.33
Dry year	South Platte	Foothills	0.2
	Roberts Tunnel	Marston	0.4
	Moffat	Moffat	0.4

than the others do. By combining this information with data in the annual report on the differential use of each component, we estimated the operating cost of each system component. Again, because the true operating costs are unknown, these variables are also subject to sensitivity analysis to test how they affect model results. Table A.2 shows our assumptions about operating costs.

To develop these estimates, we used data on average costs and relative system use with assumptions about relative cost differences. Our estimates are based on the following identity:

$$\sum_i \alpha_i \beta_i \text{average cost} = \text{average cost},$$

where index i refers to the system component (South Platte, Roberts Tunnel, and Moffat), α_i is the share of system component i of total use for a particular function and β_i is the cost multiplier for system component i, and average cost is the average cost calculated for a particular function using data in the 2005 annual report. The operating cost for an individual system component is

$$\text{operating cost} = \beta_i \text{average cost}.$$

The operating cost for a system component is the product of the cost multiplier and the average cost for a function. Based on the identity above, we assume cost multipliers for two of the three system components, and the final multiplier is constrained to ensure that the average system costs equal the average cost calculated from the data. Tables A.3, A.4, A.5, A.6, and A.7 show the average costs, α values, and β values used in the analysis.

The average cost estimates are calculated from the 2006 budget. We divided the 2005 total operating costs for each system function by the amount of water provided in each function. This figure gives an average cost for the entire system for each function. The α values use data in the 2005 annual report, which give the amount of water provided for each function

Table A.2
Operating-Cost Initial Assumptions

Function	System Component	Operating Cost ($/million gallons)
Supply	South Platte	51
	Roberts Tunnel	102
	Moffat	59
Treatment	Foothills Plant	136
	Marston Plant	271
	Moffat Plant	233
Conveyance	South Platte	291
	Roberts Tunnel	581
	Moffat	499

by system component. We used this information to calculate each system component's share of the system total for each use. The cost multipliers are made by assumption but informed by some qualitative information in agency reports. The IRP states that the Foothills Treatment Plant is the baseload plant due to its lower operating costs. Based on this information, we use a lower cost multiplier for this system component and the South Platte system. We use a high cost multiplier for the Marston and Roberts Tunnel systems, and the final multipliers are calculated using the system shares and average cost information. These cost multipliers are highly uncertain because of the limited information available. For this reason, we used a sensitivity analysis to test how they would affect the model results.

Table A.3
Components of Operating-Cost Assumptions: Average Cost Estimates

Function	Average Cost ($/million gals)
Raw-water supply	68
Water treatment	181
Delivery	388

SOURCE: Denver Water (2006a).

Table A.4
Components of Operating-Cost Assumptions: Alpha Values as Share of Total

System Component	Raw Water	Delivery
South Platte	0.49	0.59
Roberts Tunnel	0.30	0.14
Moffat Tunnel	0.20	0.27

SOURCE: Denver Water (2005).

Table A.5
Components of Operating-Cost Assumptions: Alpha Values as Share of Total

System Component	Treatment
Foothills	0.59
Marston	0.14
Moffat	0.27

SOURCE: Denver Water (2005).

Table A.6
Components of Operating-Cost Assumptions: Beta Values as Cost Multiplier

System Component	Raw Water	Delivery
South Platte	0.75	0.75
Roberts Tunnel	1.5	1.5
Moffat Tunnel	0.87	1.29

Table A.7
Components of Operating-Cost Assumptions: Beta Values as Cost Multiplier

System Component	Treatment
Foothills	0.75
Marston	1.5
Moffat	1.29

Future Supply Projects

Figure A.1 from Denver Water's IRP shows projected demand and supply to 2050.

Figure A.1 shows that Denver Water projects a supply shortfall beginning in 2016 and that, by 2030, the shortfall is 34 taf. By 2050, the shortfall is 78 taf. Denver Water currently plans 16 taf of conservation by 2030 and 29 taf by 2050. The remaining shortfalls of 18 taf and 49 taf will be met by new supply projects. The current proposal for the Moffat Collection System is to address the near-term need of 18 taf by 2030. The longer-term supply projects are unspecified.

In the AC model, we treated these plans as two future supply projects. The first is an 18-taf project that will be online by 2015, and the second is a 31-taf project online by 2030. We assumed that the near-term project could be deferred but not downsized because demand projections continue to increase enough that at least 18 taf will be needed. We treated the second project as deferrable also because, under most assumptions for demand, Denver Water will still need a second supply project.

Figure A.1
Denver Water Supply and Demand Forecast to 2050

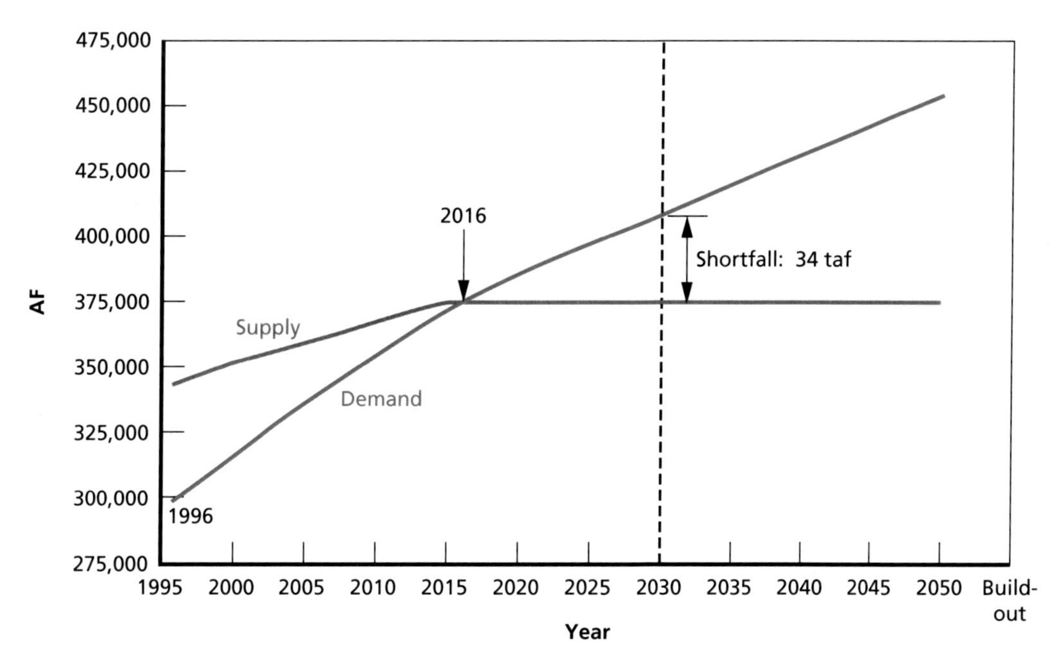

SOURCE: Adapted from Denver Water (2004).
NOTE: taf = thousand acre-feet.
RAND *TR504-A.1*

We estimated the cost of these future projects from two sources. Denver Water's 2006 annual budget projects 10-year program costs of the Moffat Project at $243.9 million but states that the total project cost is still unknown. As a lower bound, the cost translates into a supply cost of $13,548 per AF annual capacity. A second cost estimate comes from the Bureau of Reclamation and Colorado Springs Utilities' *Southern Delivery System Environmental Impact Statement* (2004). An appendix in this document reviewed cost estimates for current water-development projects in Colorado's Front Range. The review assessed cost estimates for six supply projects, including the Moffat Collection System Project. Of these, only the Colorado–Big Thompson is completed and in use. We selected the cost range for this project for use in the model. They projected a range of costs from $17,000 to $25,000 per AF of firm water yield and have used these estimates for the unspecified long-term supply project.

Modifications to the Avoided-Cost Model

Supply and Demand Forecasting Model

The original version of the AC model asks the user to specify a demand schedule and online dates of new supply projects. The model takes these values as given and does not consider uncertainty in future demand and supply. We modified the model to permit the examination of uncertainty in these factors.

We first added a factor that adjusts demand growth throughout the planning horizon, which we used to examine uncertainty in future demand. We developed a supply forecast using Denver Water's projections (see Figure 2.2). It projects a consistent increase in water supply until 2015, when water supply levels off at 375 taf. We added an adjustment factor that allowed us to vary future water supply by a percentage.

Using this supply and demand forecast, we then calculated when Denver Water would need the two future supply projects—when demand exceeds supply, the model adds a new supply project. We have maintained the project sizes anticipated by Denver Water (18 taf for Moffat Project and 31 taf for the second project). When we varied future demand and supply, the online dates for the new projects adjusted accordingly.

These uncertain factors have important implications for the LR avoided costs. The demand and supply interaction has two opposing effects on avoided costs. Shifting online dates affects the PV of avoided costs, and the demand slope affects the deferral period used in LR avoided-cost calculations as well. When demand decreases, the deferral period is larger and the LR avoided-cost increase. However, the demand decrease also pushes the online date of new projects further into the future, and the PV of the avoided costs declines. Appendix C provides a brief exploration of these effects.

Present-Value Calculation

The AC and EB models report results for each year of the planning period. We calculated the PV of these future streams of benefits to make consistent comparisons across the different categories of benefits and to estimates of utility efficiency-program costs. The nominal discount rate applied in this analysis is 6 percent. Formally, the calculation is

$$PV = \sum_t \frac{benefit_t}{(1+r)^t},$$

where t indexes the year in the planning horizon, $benefit_t$ is the value of the benefit in year t, and r is the discount rate.

APPENDIX B

Environmental-Benefit Modeling

We modified the CUWCC EB model to reflect the benefits to water-use efficiency in the Denver Water service area. We consider four environmental and recreational services: riparian and wetland habitat, river fishing, air quality, and nonangling river recreation (e.g., river rafting).

Riparian Habitat

Riparian habitat and wetlands provide various ecological benefits when maintained through sufficient water flow. The extraction and use of water by a utility can reduce flow and impact these ecosystems. The EB model bases estimates of the relationship between water availability and area of riparian habitat on an analysis of consumptive water needs of the dominant riparian and wetland species. The environmental-benefit report suggested that this methodology is preliminary and experimental. Although we adopted this methodology for the Denver Water case study, we recognize its limitations and suggest that this methodology be a candidate for further improvement.

The methodology can be disaggregated into five steps.

Step 1: Estimate Potential Area of Riparian Habitats and Wetlands

Even though riparian habitats and wetlands cover a very small percentage of the land area in Colorado, they directly or indirectly support more than half of the wildlife species in the state (Culver, 2001; Colorado Division of Wildlife, undated). As there is no clear distinction between riparian habitat and wetlands in Colorado, we estimated a river and stream buffer area that might include a combination of riparian habitat and wetlands. Acknowledging that the fraction of wetlands and riparian habitats in the estimated buffer zone is uncertain, we estimated the value of this buffer area as a function of values associated with wetlands and riparian habitats.

To estimate the potential area of riparian habitat and wetlands, we first estimated the area of riparian habitat that would exist under natural conditions. Our estimates are based on the length of river miles in each water system and an average buffer zone width, which is assumed to be between 25 and 75 feet.

We estimated the length of river miles per water system from the map available in the Denver Water annual financial report (Denver Water, 2005) and USGS maps.[1] We consider

[1] Estimated as the length of the river between Denver, Colorado, and Kersey, Colorado (USGS, 2002).

the major rivers per water system—South Platte River in the South Platte water system, Blue River in the Roberts Tunnel water system and Fraser River in the Moffat water system. For the South Platte River basin urban-agricultural setting, we considered the length of the section between Denver and Kersey to be most impacted by Denver Water Utility water-conservation decisions. The length of this section is approximately 50 miles. For the Blue River, we considered the length of the section of the river after the Dillon Reservoir to be most impacted by Denver Water Utility water-conservation decisions. The length of this section is approximately 40 miles. For the Fraser River, we considered the length of the section of the river after Winter Park to be most impacted by Denver Water Utility water-conservation decisions. The length of this section of the river is approximately 25 miles.

Using the total river miles estimated in each water system, assuming an average buffer zone of the width of 50 feet, we calculated the total potential riparian-habitat acreage. For our uncertainty analysis, we varied the buffer width between 25 and 75 feet. Table B.1 presents the total river miles estimated in each water system, the average buffer-zone width, and the estimated total potential riparian-habitat acreage.

Step 2: Estimate Consumptive Water Needs

Following the methodology of the environmental-benefit report, we estimated the consumptive water needs of a reference plant in riparian habitats, based on ET data obtained from Denver Water (undated[b]). Because ET refers to the loss of soil water through evaporation and transpiration by plant stomata, estimating consumptive water needs by ET could lead to an underestimation. ET is calculated for reference plants, usually grass or alfalfa, and we assumed a cool-season grass that is 5 inches tall for our ET reference (Colorado ET, undated). We averaged the ET estimates for three locations: Moffat weather station, Marston weather station, and the 56th Avenue pump station. Table B.2 shows monthly average consumptive water use, as estimated by ET. As presented in this table, 4.32 feet per year is required to replace water loss to ET from 1 acre of vegetation.

Step 3: Make Monthly and Seasonal Adjustments

To account for monthly and seasonal weather variations (e.g., temperature, humidity), we considered the percentage of the total acres saved as a result of water conservation in a month to be equal to the percentage of the total annualized consumptive water use for that month. Using this assumption, we calculated the number of acres sustained by 1 AF water per year in each month. This calculation is shown in Table B.3. The numbers in the rightmost column of

Table B.1
Potential Riparian Habitat and Wetland Acreage

System	Water Source	Length (miles)	Average Buffer-Zone Width (feet)	Total Potential Riparian Habitat and Wetland Acreage (acres)
South Platte	South Platte River	50	50	303
Roberts Tunnel	Blue River	40	50	242
Moffat	Fraser River	25	50	152

SOURCES: Denver Water (2005), USGS (2002).

Table B.2
Average Consumptive Water Use (estimated by ET) (feet)

Month	Moffat Water Use	Marston Water Use	56th Avenue Water Use	Average Water Use
January	0.19	0.20	0.13	0.17
February	0.18	0.19	0.18	0.18
March	0.26	0.28	0.28	0.27
April	0.42	0.45	0.46	0.44
May	0.50	0.55	0.58	0.54
June	0.58	0.63	0.70	0.64
July	0.52	0.58	0.63	0.57
August	0.44	0.49	0.53	0.48
September	0.32	0.34	0.27	0.31
October	0.23	0.24	0.22	0.23
November	0.38	0.42	0.20	0.34
December	0.14	0.14	0.14	0.14
Total	4.15	4.49	4.31	4.32

SOURCE: Denver Water (undated[b]).

NOTE: Monthly ET corresponds to 2005 or 2006 data, whichever were available.

Table B.3
Monthly Amount of Riparian Habitat and Wetlands Sustained by 1 Acre-Foot Water Saved per Year

Month	Average Consumptive Water Use (ft)	Percentage of Total Consumptive Water Use (%)	Acres Sustained by 1 AF Water Saved per Year
January	0.17	4	0.0092
February	0.18	4	0.0097
March	0.27	6	0.0147
April	0.44	10	0.0237
May	0.54	13	0.0291
June	0.64	15	0.0341
July	0.57	13	0.0308
August	0.48	11	0.0260
September	0.31	7	0.0167
October	0.23	5	0.0122
November	0.34	8	0.0180
December	0.14	3	0.0075
Total	4.32	100	0.2316

Table B.3 are multiplied by the input water savings to determine the number of acres maintained on average over a year.

Step 4: Calculate Attribution Factor

After estimating the ET or consumptive water use, we estimated the fraction of the conserved water that contributes to riparian consumptive use—the attribution factor. To do so, the total potential riparian water use per year (column C in Table B.4) is calculated first by multiplying consumptive water use (column A in Table B.4) and potential riparian area (column B in Table B.4).

Next, the average total river flow in each water system is estimated based on data obtained from USGS (undated). The average annual discharge for Blue River below Dillon Reservoir is 200.8 cubic feet per second (cfs) or 145.1 taf per year. The average annual discharge for Fraser River at Winter Park is 18.2 cfs or 13.15 taf per year. The average annual discharge for South Platte River at Denver is 352 cfs or 254.3 taf per year (Table B.4).

Finally, the attribution factor is calculated by dividing the total potential riparian water use per year by the total river flow in each water system. The results are presented in column E of Table B.4.

Step 5: Calculate Results

Finally, the number of AF sustained as a result of water conservation is calculated by multiplying the amount of water saved by the attribution factor by the number of acres sustained by 1 AF water saved per year. The annual results are presented in Table B.5. The EB model then combines the monthly impact data with monthly water savings due to efficiency to estimate the total area of wetlands supported by the specified amount of conservation.

Table B.4
Estimated Attribution Factor per Water System

	A	B	C=A*B	D	E=C/D
System	Water Use (ft/year)	Potential Area (1,000 acres)	Potential ETAW[a] (taf/year)	Flow (taf/year)	Attribution Factor
South Platte	4.32	303	1,310	254	5.14
Roberts Tunnel	4.32	242	1,050	145	7.21
Moffat	4.32	152	650	13.2	49.74

SOURCES: Calculations based on Denver Water (2005, 2006b), USGS (undated, 2002).

[a] ETAW = evapotranspiration of applied water; it represents the total potential consumptive water use.

Table B.5
Environmental Impacts on Riparian Habitat and Wetlands, South Platte System

System	Acres Sustained by 1 AF per Year	Attribution Factor	Acres Sustained by 1 AF per Year
South Platte	0.2316	0.005	0.00119
Roberts Tunnel	0.2316	0.007	0.00167
Moffat	0.2316	0.050	0.01152

Fisheries

The environmental-benefit report estimates the economic benefit of California salmon fisheries by multiplying the estimated impact of water conservation on the number of fish caught per AF times the average value of fish caught in dollars. Due to limited data available for Colorado, we could not estimate the environmental impact of water conservation on fish populations. Instead, we used two estimates of the economic impact of fishing in Colorado by the American Sportfishing Association for 2001 and 2003 (American Sportfishing Association, undated[a], undated[b]). These estimates are $1.54 billion and $1.38 billion for 2001 and 2003 respectively.

We estimated the contribution of each of the rivers to fishing in Colorado as a function of their length. The result is presented in Table B.6. We reflected uncertainty associated with this estimate and thus varied the South Platte contribution between 0 and 10 percent in our analysis.

Next, we estimated how much of this change in economic impact is attributable to changes in river flow for each of the three rivers. We then estimated the relationship between differences in economic activity between 2001 and 2003 and annual river flow for the same years. The results are presented in Table B.7. Note that the Fraser River annual flow change between 2001 and 2003 is opposite that of the statewide economic impact; we thus conclude that the data do not support estimating the impact of use for the Fraser River.

Air Emissions

The emission calculation in this application of the EB model closely follows the original version of the model. However, where possible, we modified parameters for the Rocky Mountain region. When information specific to Denver Water was unavailable, we used the default

Table B.6
Contribution to Economic Impact on Fishing

River	Length of River Affected by Denver Water Use (miles)	Contribution to Fishing (%)
South Platte	350	5
Blue	75	1
Fraser	40	1

Table B.7
Differences in Economic Impact on River Flow and Fishing Between 2003 and 2001 and the Associated Estimated Impact of Water Use Under Base-Case Assumptions

River	Annual River Flow, 2001 (taf)	Annual River Flow, 2003 (taf)	Annual River-Flow Difference (taf)	Difference in Economic Impact ($ thousands)	Impact of Use ($/af)
South Platte	170.3	162.3	7.9	4,200	529
Blue River	93.7	54.8	38.9	900	23
Fraser River	8.9	10.8	−1.9	480	—

values in the model or developed a range of possible values based on literature values and applied exploratory analysis to determine how sensitive the results were to the initial assumptions. The basic calculation and range of values used in the analysis are outlined below.

This calculation monetizes the benefits of reducing NO_X emissions. Most water utilities use electric motors to pump some of their water supply to customers, and producing the electricity used in these motors results in several types of air pollution, including NO_X. NO_X is a pollutant of great concern in many locations because it is a precursor to smog, which can have severe health effects. By decreasing a utility's demand for pumping water, water conservation can reduce harmful pollution emissions.

The model calculates NO_X emission benefits by multiplying the energy demand for pumping water by the emissions released per unit of energy and the cost of reducing NO_X emissions. The calculation has the following form:

$$\begin{aligned}&\text{energy demand rate}[\text{kwh electricity}/\text{million gallons water}]\\&\times\text{emission intensity}[\text{lbs } NO_X \text{ emitted}/\text{kwh electricity}]\\&\times\text{cost of emissions}[\$/\text{lb of } NO_X].\end{aligned}$$

For this calculation, the user enters a utility-specific energy-demand rate, the emission intensity for the local electricity supply, and the emission cost. The default values in the spreadsheet refer to California-specific figures. In this application, we used Colorado-specific values as much as possible and then relied on the default values to define a range used for exploratory analysis. The model also allows the user to specify different demand rates for peak and off-peak seasons.

We calculated the emission intensity using projections from the EIA's *Annual Energy Outlook 2007* (EIA, 2007). In the supplemental tables to that report, EIA projected electricity generation and pollution emissions by Federal Energy Regulatory Commission (FERC) reliability region. We took 2007 projections for the Rocky Mountain region and calculated the pounds of NO_X emissions per kwh of electricity (0.002 lbs per kwh) (EIA, 2007).

We relied on model default values for the energy-demand rate because we were unable to obtain the needed information from the utility. For peak season, we assumed energy demand of 649 kwh per million gallons of water and, in the off-peak season, energy demand of 325 kwh per million gallons of water.

For the cost of NO_X emissions, we assumed a range of values based on two estimates. The default value in the model is the average cost of a NO_X emission permit traded in California (\$6,000 per ton of NO_X). We used this as an upper bound, and, for the lower bound, we relied on a recent estimate of the marginal cost of controlling NO_X. In the technical analysis for the Clean Air Interstate Rule (EPA, 2007), the EPA estimated the marginal cost of controlling NO_X at \$1,250 per ton (EPA, 2005).

The EB model produces annual estimates of environmental benefits (in terms of avoided environmental costs) during the peak and off-peak seasons for each year in the planning horizon. The model calculates these values in both nominal and real terms. To evaluate the effect of these uncertainties on the overall value of water-use efficiency, we calculated the PV of all future avoided costs.

Nonangling River Recreation

We estimated the economic benefit of river rafting and impact of water use on this benefit using a similar methodology to those used by the environmental-benefit report. We first assumed that the number of users of a water body is positively related to its available water. Next, we assumed that the value of a river-recreation day is equal to the total economic impact of river rafting in Colorado for 2005 divided by the total number of user-days in Colorado. Finally, we assumed that river benefits occur only during the rafting season—May through September.

We obtained data on commercial user-days in the state of Colorado from 1988 to 2006 (Colorado River Outfitters Association, undated). Because visitor data are available only for the Blue and South Platte rivers, we excluded the Fraser River from our analysis. We obtained data on average annual discharge of these two rivers from 1988 to 2006 from the USGS National Water Information System (USGS, undated). We then used linear regression to estimate the relationship between the number of visitors and river flow, controlling for time trends in visitors. The regression coefficient for river flow is used as an estimate of the marginal impact of river flow on the number of visitors.

Tables B.8 and B.9 present the rafting visitor data and river flow data for 1988 to 2006. Table B.10 shows the results from the linear regression analysis. As can be seen, there is a positive statistical relation between the water flow and commercial user-days in the Blue River (also shown graphically in Figure B.1). There is no statistically significant relationship between the water flow and commercial user-days in the South Platte River.

Table B.8
Commercial User-Days in Colorado for the Blue and South Platte Rivers (1988–2006)

Year	Blue River (user-days)	South Platte River (user-days)
1988	2,138	—
1989	1,400	—
1990	1,928	—
1991	2,288	5,000
1992	2,173	5,000
1993	4,129	700
1994	416	101
1995	9,338	816
1996	5,854	1,112
1997	5,361	3,137
1998	1,300	3,650
1999	2,100	1,306
2000	2,347	2,035
2001	14	2,055
2002	0	453
2003	264	935
2004	788	836
2005	1,212	901
2006	760	655

SOURCE: Colorado River Outfitters Association (undated).

Table B.9
Average Annual Discharge of Blue River and South Platte River (1988–2006)

Year	Blue River Flow (taf/year)	South Platte River Flow (taf/year)
1988	131	252
1989	95	180
1990	107	179
1991	116	178
1992	94	176
1993	136	143
1994	92	146
1995	262	498
1996	298	155
1997	257	232
1998	151	375
1999	217	398
2000	129	185
2001	94	170
2002	53	87
2003	55	162
2004	54	159
2005	74	233
2006	106	158

SOURCE: USGS (undated).

Table B.10
Linear Regression Results of Number of River-Rafting Visitor Days on Annual River Flow and Time for the Blue and South Platte Rivers

River	Regression Coefficient for Annual River Flow	Standard Error	Linear Regression (R^2)
Blue River	0.026[a]	0.0043	0.725
South Platte River	0.00038	0.0034	0.25

[a] Indicates statistical significance beyond the 95 percent confidence level.

Figure B.1
Blue River Rafting Commercial User-Days Against Average Annual Flow

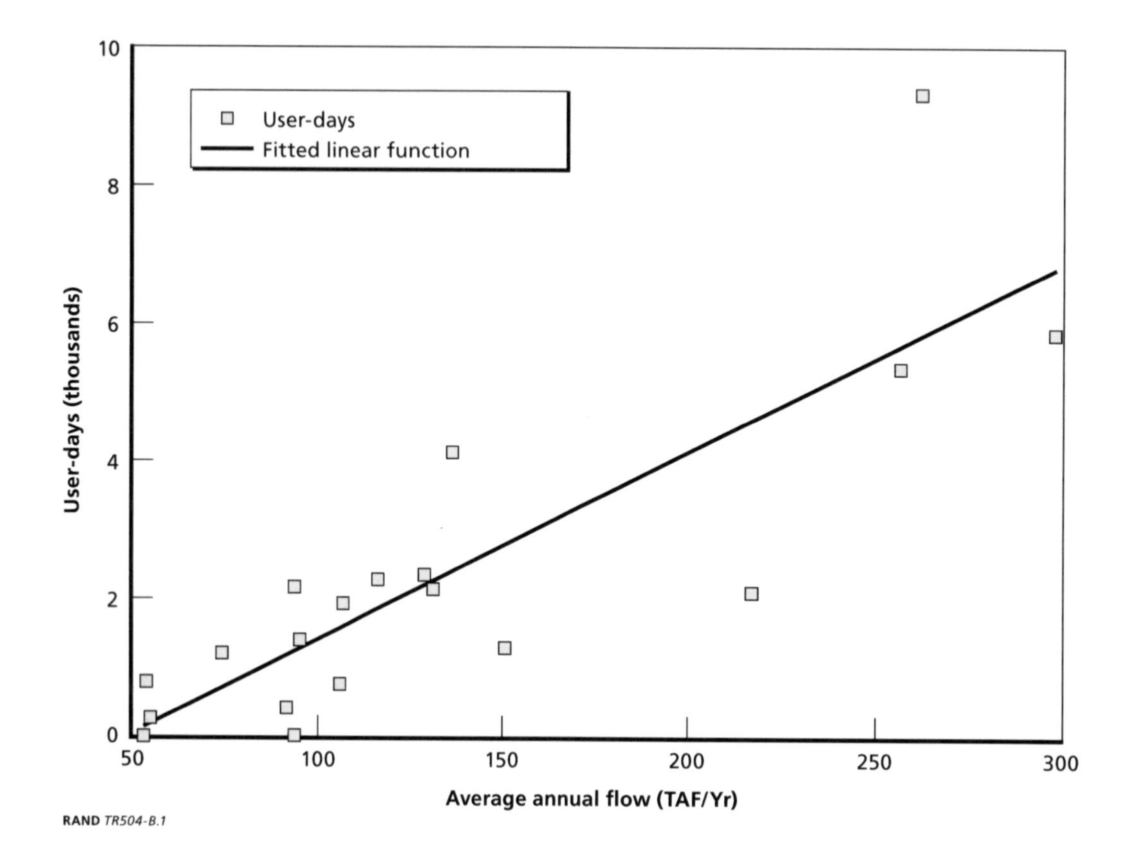

RAND *TR504-B.1*

APPENDIX C

Impact of Supply and Demand Changes on Long-Run Avoided Costs

Changes in demand and supply projections have two effects on the valuation of LR avoided costs. The first is related to the extent to which future benefits are discounted back to the present. If future projects are deferred to later periods because of lower demand growth (as in the example in Chapter Three, Figure 3.2), the avoided costs occur later in the simulation and therefore are more heavily discounted when considering the PV of the avoided costs. Lower demand growth, however, also leads to larger marginal deferral periods, leading to larger avoided-cost benefits. Note that these two effects work in opposite directions.

To better understand how the particulars of the Denver Water demand and supply projections are influenced by these effects, Figure C.1 shows how the PV of the LR avoided costs changes as a result of varying demand (from –10 percent to +10 percent) and changes in supply (from –20 percent to 0 percent). Supply decreases lead to lower PV estimates of the LR avoided costs, because the effect that a decreasing supply has on the deferral rate is greater than the effect that earlier benefits have (due to an earlier online project date). Similarly, increasing demand also leads to decreasing PV of LR avoided costs (the solid line between rate changes of 0 and 1). Interestingly, a decrease in demand also leads to decreased PV of LR avoided costs. In this case, the discounting effect due to a later online date is larger than the extended deferral period for slower-growing demand.

Figure C.1
Present Value of Long-Run Avoided Cost Under Changing Demand and Supply Forecasts

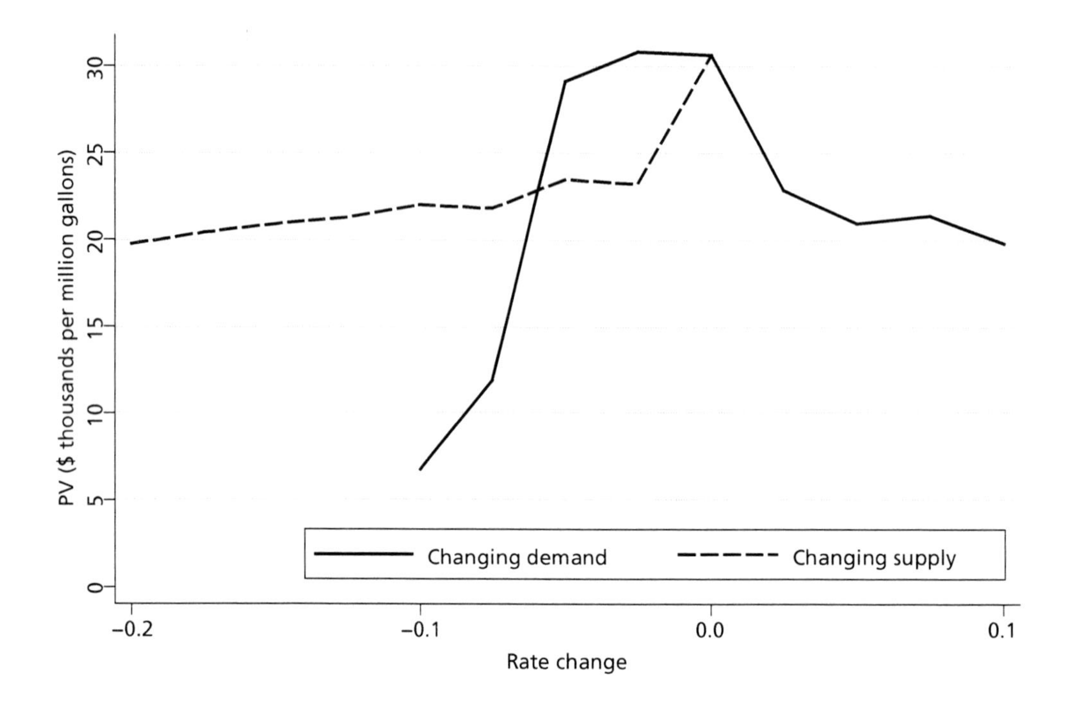

APPENDIX D

Efficiency-Program Cost Estimates

This appendix provides the raw data and calculations for the efficiency-program water savings and cost evaluation shown in Chapter Four. Table D.1 provides the raw data on the costs and savings for Denver Water proposed efficiency programs obtained from Denver Water's draft proposed 10-year conservation plan (Denver Water, 2006b). Using these data, we calculated the PV of the program cost when it is extended over the entire planning period (2007–2050) (Table D.1, seventh column). This calculation assumes that the average annual cost remains constant over time, and we calculated this cost over the entire planning horizon to ensure consistency with the benefit measures. The last two columns show the annual water savings and average cost. Note that the programs are ordered by increasing average cost.

Table D.1
Efficiency Program Measures in Denver Water's Proposed 10-Year Conservation Plan

Type	Program Measure	Projected 10-Year Denver Water Spending ($)	Projected 10-Year Consumer Spending ($)	Projected 10-Year Total Spending ($)	Annual Cost ($/yr)	PV Costs ($)[a]	Water Savings (mgal/yr)[a]	Average Cost ($/mgal)[a]
R	Low-flow–urinal requirement	31,986	45,694	77,680	7,768	119,497	98	1,223
E	Cooling-tower monitoring[b]	644,753	229,208	873,961	87,396	1,344,430	524	2,564
R/I	ET controller rebate[b]	152,667	254,446	407,113	40,711	626,269	234	2,682
E	Conservation-education program	4,574,005	0	4,574,005	457,401	7,036,275	963	7,310
R/I	Low-flow toilet	133,121	120,023	253,144	25,314	389,416	41	9,488
E	Car-wash certifications	53,379	53,379	106,758	10,676	164,228	15	10,960
E	Irrigation classes and seminars[b]	1,821,441	1,915,345	3,736,786	373,679	5,748,366	474	12,121
R	Irrigation-meter requirement[b]	52,914	2,116,577	2,169,491	216,949	3,337,367	205	16,315
R/I	Wireless rainfall-sensor rebate[b]	545,151	218,061	763,212	76,321	1,174,063	54	21,713
E	Conservation kiosks	1,201,088	0	1,201,088	120,109	1,847,656	85	21,817
R	Time-of-sale retrofit of toilets, showerheads, and faucets	19,528,618	29,584,243	49,112,861	4,911,286	75,551,208	2,278	33,168
E	Multifamily residential audit program	123,697	363,813	487,510	48,751	749,946	22	33,367

Table D.1—Continued

Type	Program Measure	Projected 10-Year Denver Water Spending ($)	Projected 10-Year Consumer Spending ($)	Projected 10-Year Total Spending ($)	Annual Cost ($/yr)	PV Costs ($)[a]	Water Savings (mgal/yr)[a]	Average Cost ($/mgal)[a]
R	Water-efficiency rating for new customers	8,289,287	19,286,808	27,576,095	2,757,610	42,420,809	1,236	34,335
R/I	High-efficiency–toilet rebate	2,789,289	1,503,739	4,293,028	429,303	6,604,043	134	49,450
R/I	Natural-area conversion for large landscapes[b]	22,581,652	0	22,581,652	2,258,165	34,737,766	563	61,680
R/I	Commercial and industrial incentives	18,300,314	58,014,980	76,315,294	7,631,529	117,397,206	1,515	77,491
R/I	Clothing-washer rebate	12,241,670	12,359,379	24,601,049	2,460,105	37,844,241	438	86,445
E	Irrigation checkups for large irrigators[b]	4,180,280	31,625,674	35,805,954	3,580,595	55,080,951	619	88,999
R/I	Public-housing retrofits	3,098,186	442,347	3,540,533	354,053	5,446,466	51	105,827
R/I	Irrigation-efficiency incentives[b]	22,074,512	87,016,763	109,091,275	10,909,128	167,817,094	675	248,528
E	Xeriscape planning and design[b]	187,965	10,630,165	10,818,130	1,081,813	16,641,726	21	810,954

SOURCE: Denver Water (2006b).

[a] Computed by the authors.

[b] Likely to concentrate savings during peak months.

NOTE: E = educational. R/I = rebate or incentive. R = regulatory.

References

A&N Technical Services Inc., and Gary Fiske and Associates, *Water Utility Direct Avoided Costs from Water Use Efficiency*, Sacramento, Calif.: California Urban Water Conservation Council, January 2006. As of October 26, 2007:
http://www.cuwcc.org/Uploads/committee/Avoided_Costs_Draft_Final_06-09-18.pdf

American Rivers, *America's Most Endangered Rivers of 2005: Ten Rivers Reaching the Crossroads in the Next 12 Months*, Washington, D.C., undated. As of October 29, 2007:
http://www.americanrivers.org/site/DocServer/AR_MER_2005.pdf?docID=1261

American Sportfishing Association, "Data and Statistics: Sales and Economic Trends: 2001 State by State Economic Impacts—All Types of Fishing: Freshwater, by State," undated Web page (a). As of October 29, 2007:
http://www.asafishing.org/asa/statistics/saleco_trends/state_reports_freshwater.html

———, "Data and Statistics: Sales and Economic Trends: 2003 Estimated Economic Impacts by State," undated Web page (b). As of October 29, 2007:
http://www.asafishing.org/asa/statistics/saleco_trends/state_allfish_2003.html

Bankes, Steve, "Exploratory Modeling for Policy Analysis," *Operations Research*, Vol. 41, No. 3, May 1993, pp. 435–449.

BBC Research and Consulting, and Colorado Division of Wildlife, *The Economic Impacts of Hunting, Fishing and Wildlife Watching: Final Report*, Denver, Colo.: BBC Research and Consulting, 2004. As of October 29, 2007:
http://www.cde.state.co.us/artemis/nr6/nr62f522004internet.pdf

Bureau of Reclamation, and Colorado Springs Utilities, *Southern Delivery System Environmental Impact Statement: Review of Cost Estimates for Current Water Development Projects in the Front Range*, 2004.

CALFED—*see* FED Bay-Delta Program and California Bay-Delta Authority.

CALFED Bay-Delta Program, and California Bay-Delta Authority, *Water Use Efficiency Comprehensive Evaluation: Final Report*, Sacramento, Calif.: CALFED Bay-Delta Authority, 2006. As of October 26, 2007:
http://digitalarchive.oclc.org/request?id%3Doclcnum%3A144594198

Chappell, Linda M., *Regulatory Impact Analysis for the Final Clean Air Interstate Rule*, Research Triangle Park, N.C.: U.S. Environmental Protection Agency, Office of Air and Radiation, Air Quality Strategies and Standards Division, Emission, Monitoring, and Analysis Division, Clean Air Markets Division, 2005. As of October 30, 2007:
http://www.epa.gov/air/interstateairquality/pdfs/finaltech08.pdf

Christopherson, Karen, "Don't Forget the Fraser," *Colorado Fishing Network*, undated Web page (a). As of October 29, 2007:
http://www.coloradofishing.net/ft_fraser.htm

———, "Species of Fish in Colorado," *Colorado Fishing Network*, undated Web page (b). As of October 29, 2007:
http://www.coloradofishing.net/species.htm

Colorado Division of Wildlife, "Wetland Versus Riparian," undated Web page. As of October 29, 2007:
http://ndis1.nrel.colostate.edu/riparian/Ripwetdef.htm

Colorado ET, undated homepage. As of October 31, 2007:
http://www.coloradoet.org/

Colorado River Outfitters Association, "Executive Summary: Commercial River Use in Colorado," undated. As of October 30, 2007:
http://www.croa.org/pdf/2006_Commercial_Rafting_Use_Report.pdf

Coughlin, Katie, Chris Bolduc, Peter Chan, Camilla Dunham-Whitehead, Robert van Buskirk, W. Michael Hanemann, and Katie Shulte Joung, *Valuing the Environmental Benefits of Urban Water Conservation: Final Report*, Sacramento, Calif.: California Urban Water Conservation Council, March 27, 2006. As of October 26, 2007:
http://www.cuwcc.org/Uploads/committee/EB_Final_Report_06-03-27.pdf

Culver, Denise, "Status of Colorado Natural Heritage Program Wetlands Project," *Green Line Online*, Vol. 12, No. 4, Winter 2001. As of October 29, 2007:
http://coloradoriparian.org/GreenLine/V12-4/Status.html

Cutthroat Anglers, "River and Lake Information: Blue River," undated Web page. As of October 29, 2007:
http://www.fishcolorado.com/river-info/blue.html

Denver Water, "Dillon Reservoir," undated Web page (a). As of October 29, 2007:
http://www.denverwater.org/recreation/dillon.html

———, "Reservoir Levels and More," undated Web page (b). As of October 31, 2007:
http://www.denverwater.org/reslevelsmore/reslevelsmoreframe.html

———, *Water for Tomorrow: The History, Results, Projections, and Update of the Integrated Resource Plan*, Denver, Colo.: Denver Water, 2002.

———, *Purpose and Need Statement for the Moffat Collection System Project*, Denver, Colo.: Denver Water, April 2004. As of October 26, 2007:
http://www.denverwater.org/pdfs/finalpurposeneed40704.pdf

———, *Denver Water 2005 Annual Report*, Denver, Colo.: Denver Water, 2005.

———, *2006 Budget*, Denver, Colo.: Denver Water, March 1, 2006a. As of October 26, 2007:
http://www.denverwater.org/financialinfo/budgetreport/BudgetBook2006.pdf

———, *DRAFT—Proposed 10-Year Conservation Plan (July 6, 2006)*, Denver, Colo.: Denver Water, 2006b.

EIA—*see* Energy Information Administration.

Energy Information Administration, *Annual Energy Outlook 2007*, Washington, D.C.: U.S. Government Printing Office, DOE/EIA-0383(2007), 2007. As of October 30, 2007:
http://www.eia.doe.gov/oiaf/aeo/index.html

EPA—*see* U.S. Environmental Protection Agency.

Frederick, Shane, George Loewenstein, and Ted O'Donoghue, "Time Discounting and Time Preference: A Critical Review," *Journal of Economic Literature*, Vol. 40, No. 2, June 2002, pp. 351–401.

Gleick, Peter H., Dana Haasz, Christine Henges-Jeck, Veena Srinivasan, Gary Wolff, Katherine Kao Cushing, and Amardip Mann, *Waste Not, Want Not: The Potential for Urban Water Conservation in California*, Oakland, Calif.: Pacific Institute for Studies in Development, Environment, and Security, November 2003. As of October 26, 2007:
http://www.pacinst.org/reports/urban_usage/waste_not_want_not_full_report.pdf

Groves, David G., Jordan Fischbach, and Scot Hickey, *Evaluating the Benefits and Costs of Increased Water-Use Efficiency in Commercial Buildings*, Santa Monica, Calif.: RAND Corporation, TR-461-NAT, 2007. As of November 5, 2007:
http://www.rand.org/pubs/technical_reports/TR461/

Lempert, Robert J., Steven W. Popper, and Steven C. Bankes, *Shaping the Next One Hundred Years: New Methods for Quantitative, Long-Term Policy Analysis*, Santa Monica, Calif.: RAND Corporation, MR-1626-RPC, 2003. As of October 30, 2007:
http://www.rand.org/pubs/monograph_reports/MR1626/

Metropolitan Water District of Southern California, *Regional Urban Water Management Plan*, Los Angeles, Calif., November 2005. As of November 5, 2007:
http://www.mwdh2o.com/mwdh2o/pages/yourwater/RUWMP/RUWMP_2005.pdf

National Resource Council, *Valuing Ecosystem Services: Toward Better Environmental Decision-Making*, Washington, D.C.: National Academies Press, 2005.

Platt, Jonathan, *Economic Nonmarket Valuation of Instream Flows*, Denver, Colo.: U.S. Department of the Interior, Bureau of Reclamation, 2001.

RAQC—*see* Regional Air Quality Council.

Regional Air Quality Council, "State Implementation Plans and RAQC Reports," undated Web page. As of October 30, 2007:
http://www.raqc.org/reports/reports.htm

USACE—*see* U.S. Army Corps of Engineers.

U.S. Army Corps of Engineers, Omaha District *Scoping Summary: Moffat Collection System Project*, Cheyenne, Wyo., December 2003. As of October 30, 2007:
http://www.denverwater.org/moffat.pdf

U.S. Census Bureau, "State and County QuickFacts: Colorado," August 31, 2007. As of October 29, 2007,:
http://quickfacts.census.gov/qfd/states/08000.html

U.S. Environmental Protection Agency, *Rule to Reduce Interstate Transport of Fine Particulate Matter and Ozone (Clean Air Interstate Rule); Revisions to Acid Rain Program; Revisions to NOX SIP Call*, 70 F.R. 25162, May 12, 2005.

———, "Clean Air Interstate Rule," last updated April 5, 2007. As of October 30, 2007:
http://www.epa.gov/cair/

U.S. Forest Service, Rocky Mountain Region, *Wild and Scenic River Study Report and Final Environmental Impact Statement: North Fork of the South Platte and the South Platte Rivers*, Pueblo, Colo.: U.S. Department of Agriculture and U.S. Forest Service, Rocky Mountain Region, 2004. As of October 29, 2007:
http://www.fs.fed.us/r2/psicc/projects/wsr/

USGS—*see* U.S. Geological Survey.

U.S. Geological Survey, "National Water Information System: Web Interface: USGS Surface-Water Annual Statistics for the Nation," undated Web page. As of October 31, 2007:
http://nwis.waterdata.usgs.gov/nwis/annual?referred_module=sw

———, "National Water-Quality Assessment (NAWQA) Program: South Platte River Basin," last modified July 1, 2002. As of October 31, 2007:
http://webserver.cr.usgs.gov/nawqa/splt/html/spbasininfops.html

Young, Robert A., *Determining the Economic Value of Water: Concepts and Methods*, Washington, D.C.: Resources for the Future, 2005.